Graphic Products

Jon Attwood

D1354793

Heinemann Educational Publishers
Halley Court, Jordan Hill, Oxford, OX2 8EJ
Part of Harcourt Education
Heinemann is the registered trademark of
Harcourt Education Limited

Text © Jon Attwood, 2002
First published in 2002

07 06 05 04
10 9 8 7 6 5 4

British Library Cataloguing in Publication Data

A catalogue record for this book is available from the British Library

ISBN 0 435 41780 0

Original illustrations ©Heinemann Educational Publishers, 2002.

Designed, typeset and illustrated by Hardlines Ltd, Charlbury, Oxford
Printed and bound in the UK by Bath ColourBooks

Acknowledgements

The author and publishers would like to thank Gillian Whitehouse and Parul Patel at Edexcel; Jackie Potton at Crofton School in Lewisham, London; the D&T department at the Ravensbourne School in Bromley, Kent; for student work – Jonathan Stiasny, Arone Thompson, Daniel Cook and a special thanks to Lisa Robinson.

The publishers would like to thank the following for permission to use copyright material:

British Airways for the logo on p.109; British Glass for the diagrams on p.27; British Standards Institution (BSI) for the BSI Kitemark and CE logo on p.92; Burton Snowboards for the brochure image on p.59; Colgate-Palmolive (UK) for the Colgate toothpaste box on p.39; E4 with kind permission for the graphics on p.57; EMI for the gorillaz.com website graphic on p.95; Gillette Management Inc. for the Duracell Powercheck® image on p.97; HSE for the diagram on p.55, adapted from the HSE diagram; Kellogg for the logo on p.108; trade mark on p.108 with kind permission of Levi Strauss UK Ltd.; the McDonald's Lozenge Logo on p.108 is owned by McDonald's Corporation and affiliates; Nike UK for the logo on p.109; Sony Computer Entertainment Europe for the Playstation Wipeout packshot on p.105; Supersine Duramark Limited for the Vinyl Film Swatch Chart on p.70; the Bassett's Dip Dab wrapper on p.56 was reproduced with kind permission of Trebor Bassett Limited, Source: *Computer Arts*, March 2001.

The publishers have made every effort to trace the copyright holders, but if they have inadvertently overlooked any, they would be pleased to make the necessary arrangements at the first opportunity.

The publishers would like to thank the following for permission to reproduce photographs:

AFP pp.86, 104; Apple p.107; Jon Atwood pp. 18 (all), 41 (top), 42, 47, 71, 125, 135, 151; Gareth Boden p.61; Corbis pp.3, 33 (suitcases), 43 (Bandai robot), 68, 91, 98; Corbis/Paul Almasy p.69; Corbis/Dean Conger p.4; Corbis/Michael Everton p.111; Corbis/James Sugar p.29; Corbis/Giovanni Reda p.32; Giles Chapman Library p.71; Denford p.88; Trevor Clifford pp. 12 (left), 28; Epson pp.60, 64; Ford pp.78, 84; Ford/Giles Chapman pp.85 (all); Chris Honeywell pp. 8, 9, 12 (right), 20, 21, 33 (right), 38, 41 (bottom), 62 (key ring); 63, 82, 94, 97, 105 (bottom), 106 (left), 111, 112, 113, 114 (all); Peter Morris pp.39, 49, 60 (high and low res images), 74, 104 (bottom), 107 (bitg), 109; Richard Opie pp.23 (both), 101; Photodisc pp.5, 22, 67, 74 (bird), 102; Quintec/Ford p.84; Science Photo Library pp.6, 90, 99, 103; SPL p.81; TechProducts pp.53 (both); TechSoft (CAD drawing) pp.57, 65 (both); Trip/Helene Rogers pp.13, 48, 62, 72, 74, 79; Yo!Sushi p.86.

Cover photographs by: Image bank (top), Haddon Davies (middle), John Birdsall (bottom)

Tel: 01865 888058 www.heinemann.co.uk

Contents

Introduction

Welcome to this GCSE student book, which has been specially written to support you as you work through your Design and Technology course. If you are following a short-course, check with your teacher to see which sections of the book you need to cover. You can find out information about the full-course and the short-course in the next couple of pages.

How to use this book

This student book will help you develop knowledge and understanding about the specialist materials area you have chosen to study within Design and Technology. It includes sections on:

- the classification and selection of materials
- preparing, processing and finishing materials
- manufacturing commercial products
- design and market influence.

As you work through each section, you will find 'Things to do', which test your understanding of what you have learned. Your teacher may ask you to undertake these tasks in class or for homework.

At the end of each section you will find a number of questions that are similar in style to the ones in the end-of-course exam. In preparation for your exam it is a good idea to put down on a single side of A4 paper all the key points about a topic. Use sub-headings or bullet point lists and diagrams to help you organize what you know. If you do this regularly throughout the course, you will find it easier to revise for the exam.

The book also includes sections that cover the coursework requirements of the full-course and the short-course. These coursework sections will guide you through all the important designing and making stages of your coursework. They explain:

- how to organize your project
- what you have to include
- how the project is marked
- what you have to do to get the best marks.

You should refer to the coursework sections as and when you need.

The GCSE Design and Technology full-course

The GCSE Design and Technology full-course builds on the experience you had of all the five materials areas at Key Stage 3:

- Food Technology
- Textiles Technology
- Resistant Materials Technology
- Graphic Products
- Systems and Control Technology.

Each of the five materials areas will provide opportunities for you to demonstrate your design and technology capability. You should therefore specialize in a materials area that best suits your particular skills and attributes.

What will I study?

Throughout the full-course you will have the opportunity to study:

- materials and components
- production processes
- industrial processes
- social, moral, ethical and environmental issues of products design
- product analysis
- designing and making processes.

The content of this student book will provide you with all the knowledge and understanding you need to cover during the full-course.

You will then apply this knowledge and understanding when designing and making a 3D product and when producing an A3 folder of design work. You should spend up to 40 hours on your coursework project, which accounts for 60 per cent of your Design and Technology course.

At the end of the full course you will be examined on your knowledge and understanding of your chosen materials area. There will be a $1\frac{1}{2}$ hour exam, worth 40 per cent of the total marks. The exam will be made up of four questions, each worth 10 per cent of the marks.

The GCSE Design and Technology short-course

The GCSE Design and Technology short-course is equivalent to half a full GCSE and will probably be delivered in half the time of the full-course. It involves the study of HALF the content of the full GCSE, and the development of HALF the amount of coursework.

The GCSE short-course allows you to work in the materials area you feel best suits your own particular skills and attributes. You can choose from:

- Food Technology
- Textiles Technology
- Resistant Materials Technology
- Graphic Products
- Systems and Control Technology.

The content of this student book will provide you with all the knowledge and understanding you need to cover during the short-course.

You will then apply this knowledge and understanding when designing and making a 3D product and when producing an A3 folder of design work. You should spend up to 20 hours on your coursework project, which accounts for 60 per cent of your Design and Technology course.

At the end of the short-course you will be examined on your knowledge and understanding of your chosen materials area. There will be a 1-hour exam, worth 40 per cent of the total marks. The exam will be made up of three questions.

Managing your own learning during the course

At GCSE level you are expected to take *some* responsibility for planning your own work and managing your own learning. The ability to do this is an essential skill at Advanced Subsidiary (AS) and Advanced GCE level. It is also highly valued by employers.

In order that you start to take some responsibility for planning your own work, you need to be very clear about what is expected of you during the course. This book aims to provide you with such information. Helpful hints include:

- Read through the whole of the introduction before you start the course so you fully understand the requirements of either the full-course or the short-course.
- Investigate the coursework sections that give you a 'flavour' of what you are expected to do.
- Check out how many marks are awarded for each of the assessment criteria. The more marks that are available, the more work you will need to achieve them.
- Discuss the coursework deadlines with your teacher so you know how much time is available for your coursework.

ICT skills

There will be opportunities during the course for you to develop your ICT capability through the use of CAD/CAM. You may have the opportunity to use:

- ICT for research and communications, such as using the Internet, e-mail, video conferencing, digital cameras and scanners
- word-processing, databases or spreadsheets for planning, recording, handling and analyzing information
- CAD software to model, prototype, test and modify your design proposals
- CAM using computer controlled equipment.

Understanding industrial and commercial practice

During your GCSE course you will have the opportunity to develop an understanding of the design and manufacture of commercial products by undertaking product analysis.

You should demonstrate your understanding of industrial practices in your designing and making activities, which could include:

- developing design briefs and specifications
- using market research
- modelling and prototyping prior to manufacture
- producing a working schedule that shows how the product is manufactured
- making a high quality product that matches the design proposal
- testing and evaluating your product against the specification to provide feedback on its performance and fitness-for-purpose.

You should also use the appropriate technical words to describe your work. Many of these words are to be found in this book. When the words first appear they are in **bold**. This means that you can look up their meaning in the glossary that appears at the end of the book.

Section A:
The classification and selection of materials and components

Exterior of Namiki House

Paper and board

Aims

- To understand the process of making paper from wood pulp.
- To understand that paper and board is available in a range of weights, sizes and finishes.

Paper and boards are the most useful material for the production of graphic products. Wood is the primary raw material for the manufacture of paper and boards because it is widely available and relatively cheap. Other materials can be used such as cotton, straw and hemp, producing papers with different properties.

The production of woodpulp

Wood is made up of fibres that are bound together by a material called lignin. In order to make paper these fibres must be separated from one another to form a mass of individual fibres called **woodpulp**. This process is carried out at a **pulp mill** by using either **mechanical** or **chemical pulping**. Mechanical pulping is used to produce newsprint for newspapers and chemical pulping produces printing and writing papers.

Quality papers require a pulp that is bright white and will not discolour with age. It is therefore necessary to **bleach** the pulp using chlorine in order to remove all impurities such as bits of tree bark. Packaging grades such as Kraftliner – used for corrugated board – are, however, left unbleached.

The production of paper using a Fourdrinier machine

Machine-made paper

The main classification of paper refers to its method of production. Machine-made paper is the most commonly used paper as it is widely available in a range of colours, sizes and finishes for various applications including printing and art and presentation work.

The production of machine-made paper is a continuous process using **Fourdrinier** machines. Essentially, the woodpulp goes in at one end and passes through a series of rollers, pressers and dryers until eventually rolls of paper come out the other end. Some Fourdrinier machines can be about a mile long!

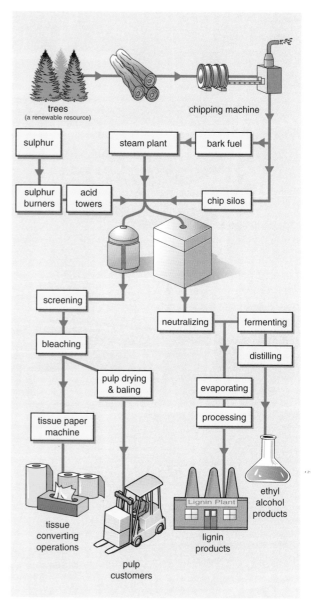

The mechanical and chemical production of woodpulp

During this process the opacity, texture, weight and colour of the paper can be determined. For example, during the final stages of production the paper is passed through a series of steel calender rollers. This operation called **calendering** increases the smoothness and gloss of the paper – the more calenders, the higher the gloss.

Texture and colour

Laid paper is produced by laying rolls of wet paper on a mesh of horizontal or vertical wires. When the paper dries out the striped impression is left. **Wove paper** is produced in a very similar way but on a mesh of woven wires.

A **watermark** can be added to paper in order to create an individual and high quality effect. A raised symbol is placed on the **dandy roll** of the Fourdrinier machine and makes the paper thinner in that shape. Therefore, when held up to the light more light can pass through the watermark than the rest of the paper.

The finish on paper refers to the way its surface has been treated. The roughest finish is called **antique** and is an **uncoated** paper. **Coated** papers include egg-shell and machine finish (MF) paper.

Coloured dyes or pigments are added to the wood-pulp during the production of paper to produce a wide range of colours.

Hand-made paper

The process of making this type of paper is very slow and expensive as each sheet has to be hand produced. It is usually used for very high quality applications such as letterheads, limited edition books and artists' paper where unique textures and patterning are important.

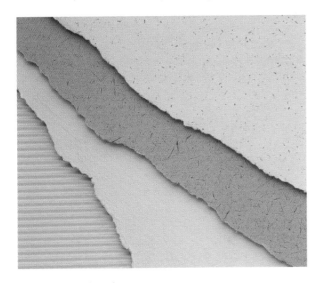

A range of hand-made papers

Common thicknesses of paper and board

Sheets	Microns
2	200
3	230
4	280
6	360
8	500
10	580
12	750

Weight

Paper is available in different thicknesses or weight which is measured in grams per square metre (gsm). Most paper used in schools will weigh 80 gsm, which is fairly thin. Card and board, on the other hand, are measured in micrometers or microns for short. Mounting board used for presentation work may be as dense as 1000 microns which is pretty thick!

When does paper become a board?

Paper usually becomes a board when it is greater than 220 gsm and more often than not is made from more than one ply (sheet). The thickness of card and board can be gauged by the number of plys or sheets it consists of.

Common sizes

Paper and board are available in metric 'A' sizes. A4 and A3 are the most commonly used in schools and offices (A3 being twice the size of A4). In addition to these there are many other sizes available, including 'B' sizes and the old imperial measurements.

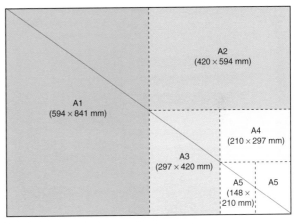

Common 'A' sizes of paper and board

■ Things to do ■

1 Try making your own hand-made paper.
2 Collect examples of laid and wove papers with watermarks to add to your notes.

Plastics

Aims

- To understand the importance of plastics.
- To understand the classification of plastics into thermoplastics and thermosets and to understand their characteristics.
- To understand the range of plastics available for packaging, sheet and block modelling.

There was a rapid growth in the use of plastics during the second half of the twentieth century. They have provided alternatives to or have completely replaced many packaging requirements previously carried out by metal, glass and cardboard.

Historical background

Plastics were first commercially introduced in the early twentieth century with a product called Bakelite. This was primarily used for the casings of electrical products such as radios as it had excellent electrical insulation properties. Soon a whole host of commercial products was being produced using different types of plastics.

An early Bakelite radio

Production of plastics

Plastics are members of a family of substances called polymers which have very large chain-like molecules. Each molecule of a polymer contains smaller units called monomers which are joined together. Polymers occur in the natural world, for example amber, animal horn and tortoiseshell. However, it is synthetic (man-made) polymers that are used for plastics.

Synthetic plastic is produced from crude oil. A system of refining and processing the basic chemicals from crude oil produces monomers. Monomers are converted into polymers which are then made into granules of plastic. The plastic granules are processed in various different ways to produce plastic products.

Thermoplastics and thermosetting plastics

Plastics can be divided into two main groups due to their specific properties once heated.

Thermoplastics

A thermoplastic is a plastic that once heated can be formed into a variety of interesting shapes using different forming techniques. The shape then remains permanent once the plastic has cooled down. The same thermoplastic can be heated, softened, shaped and cooled many times over.

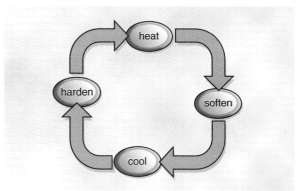

Thermoplastic heating cycle

Thermosetting plastics

The main difference between a thermoplastic and a thermosetting plastic is that once they are heated, shaped and cooled, they become permanently hard. A thermosetting plastic therefore, cannot be reheated and reshaped.

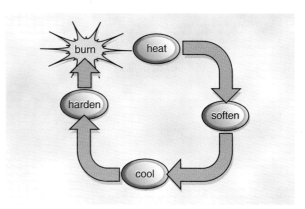

Thermoset heating cycle

Plastics in packaging

There are six main plastic materials used for the production of packaging (usually known by their initials): PET, HDPE, PVC, LDPE, PP and PS. All of these are thermoplastics.

Thermoplastics are ideal for use in packaging as they can be reheated, reshaped and can therefore be recycled. Without these plastics we could not make many of the unique and unusual designs in modern packaging.

Plastics for sheet modelling

Plastics are available in sheet form and can be used for a wide range of applications for graphic products.

Plastics for block modelling

Sheet plastics such as acrylic can be laminated (glued together) in order to produce a thick block for the production of some models. It is, however, expanded polystyrene and Styrofoam that are best used for block modelling.

Expanded polystyrene

Expanded polystyrene can be obtained free from the packaging of electrical products. Blocks can be achieved by glueing several pieces together using **PVA** glue. It is extremely easy to cut with hot wire cutters but tends to crumble when shaped or sanded.

Styrofoam

Styrofoam is a specialist modelling material for producing **concept** models. The advantages of using Styrofoam over laminated MDF are that it is easier to cut and shape. Because of its density, Styrofoam can be sanded to a very smooth finish and painted using acrylic paints.

> ### ∎ Things to do ∎
>
> 1 Collect examples of sheet plastics for your notes.
> 2 Make a concept block model of a computer mouse using styrofoam.

Material	Properties	Applications
Acrylic	Easily thermoformed, i.e. line bending and vacuum forming Range of colours available including transparent and translucent Excellent surface finish Easily joined using Tensol cement	Product prototypes and concept models Structures such as interior and architectural models Point-of-sale displays, stands and leaflet holders
Corriflute	Lightweight Easily joined using hot melt glue Rigid Range of colours available Impact and heat resistant	Structures and signage for point-of-sale displays Architectural and interior models, etc.
Foamboard	Easily cut Rigid and strong Lightweight Easily joined using mapping pins and hot melt glue Can be drawn on	Structures such as point-of-sale displays, architectural models, etc.
High impact polystyrene (HIP)	Can be vacuum formed easily Range of colours available Easily joined using liquid solvent cement	Exhibition signage Bubble packs Product prototypes and concept models Interior and architectural models
Acetate	Transparent Available for photocopiers Easily cut and scored Flexible Easily joined to a range of materials using hot melt glue	LCD display screen for prototype models Windows for architectural models Windows in nets

Sheet plastics

Wood

Aims

- To understand that woods are classified into three groups: hardwoods, softwoods and manufactured board.
- To understand the characteristics of the three groups.

Wood can be extremely useful for producing a range of graphic products from interior design models to product prototypes. It is available in a variety of shapes and sizes with each timber having its own properties.

People have always known the special properties of wood. Early humans used wood for fires, to build shelters and to hunt with. As time went on, wood was used for major engineering tasks such as architecture and shipbuilding and highly decorative uses such as furniture and intricate carvings. Every culture in all parts of the world has used wood in some way to benefit its society.

Woods can be divided into three main categories:
- **hardwoods**
- **softwoods**
- **manufactured boards.**

Both hardwoods and softwoods are produced from naturally growing trees, whereas manufactured boards are man-made using natural timber.

Hardwoods

Hardwoods are produced from broad-leaved trees whose seeds are enclosed. Examples include elm, oak, beech, balsa, mahogany and walnut.

Hardwood trees grow in warm climates such as Africa and South America and take about 100 years to reach maturity. They are usually tough and strong and provide highly decorative finishes. Because of their age and where they grow, many hardwoods are expensive to buy and may only be used in very high quality products. The exception is Balsa wood, which has been used to make models for many years as it is relatively cheap and easy to work with.

Softwoods

Softwoods are produced from cone-bearing conifers

with needle-like leaves. Examples include Scots pine (red deal), parana pine and whitewood.

As softwoods grow more quickly than hardwoods (30 years) they can be forested and replanted which means they are cheaper. Softwoods are also easier to work with and lightweight which makes them more suitable for model making.

Balancing kiwis made from farm forested Radiata Pine

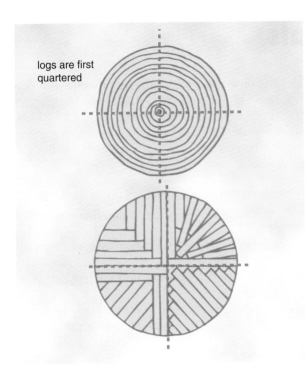

logs are first quartered

Radial sawing is used to minimize the amount of wastage

Production of hardwoods and softwoods

Once a hardwood or softwood tree has been felled (chopped down), it is transported to the timber mill where it is processed into planks or boards ready for use. The first process is called the **conversion of timber** where the tree trunk is sawn up into usable sizes using large circular saws.

The second stage involves the **seasoning** of the timber by removing all moisture by drying it in either a kiln or in the open air. This is carried out to increase the strength and stability of the timber and its resistance to decay.

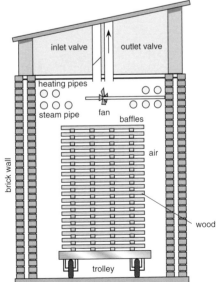

A kiln allows the timber to be seasoned quickly

Manufactured boards

Manufactured boards can be made either from thin sheets (**veneers**) of wood sandwiched together or from wood particles glued together and compressed (squashed).

The advantages of using manufactured boards are that they are available in wide boards which is not possible with natural timbers where the width depends on the width of the tree trunk.

By running the grain of the veneers at 90 degrees to one another, some boards are given added strength. They are also very much cheaper to buy than natural timber and have greater uses in model making.

There are several types of manufactured boards:

- plywood
- blockboard and laminboard
- particleboard (i.e. hardboard)
- fibreboard (i.e. medium density fibreboard or MDF).

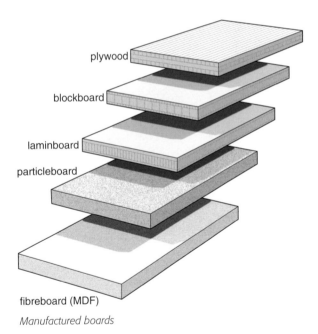

Manufactured boards

> ■ **Things to do** ■
>
> 1 Collect samples of hardwoods, softwoods and manufactured boards for your notes.
>
> 2 Investigate the standard sizes for a range of boards using a timber catalogue.

Metals

Aims

- To understand the importance of metals in packaging.
- To understand the production of aluminium and steel from ores.

Metals in packaging

The use of metal as a packaging material is extremely important in the preservation of food. Fresh food can quickly decay and become rotten which was a problem throughout history. During the early nineteenth century, Napoleon (Emperor of France) had the problem of supplying fresh food to his armies in distant countries. Nicholas Appert discovered a preserving process by firstly cooking the food and then storing it in a sealed tin canister – later shortened to tins or cans. Soldiers could now enjoy a healthier diet even when miles from home.

It wasn't until the 1920s, however, that the canning industry had real commercial success. During this time the use of fast, automated production lines producing over a thousand cans a minute was introduced. Today, over 13 000 million cans are bought every year!

These are some of the advantages of canned food:

- Cans have a long shelf life if stored in a cool, dry cupboard.
- Cans do not need to be refrigerated, which saves energy and money.
- Canning makes a wide range of foods available all year round.
- Canning and cooking preserve the food so reducing the need for artificial preservatives.

The two main types of metals used in packaging are:

- aluminium
- steel (coated with tinplate).

The basic process of canning food has not changed much since Appert came up with idea to heat food and seal it off in a can to prevent bacteria from spoiling the food – the main change is in the speed of which cans are now produced

Aluminium

Aluminium is a **pure metal** which is a naturally occurring element that is mined from beneath the land and sea. It is the most plentiful metal element in the earth's crust and is produced from the ore *bauxite*.

The production of aluminium requires large amounts of electricity due to the expensive **electrolytic process** involved. There are two main stages:

1 The production of alumina from bauxite

Once the bauxite has been mined, crushed and dried it is refined into **alumina**. This is done in two stages. First, the bauxite is dissolved in hot caustic soda and then filtered to remove impurities – aluminium oxide is produced. Secondly, the aluminium oxide is 'roasted' in a rotary kiln and a white powder is produced called alumina.

2 The production of aluminium using an electrolytic reduction cell

In the reduction cell the alumina is dissolved in molten cryolite using a steel furnace. The furnace is lined with carbon (forming a cathode) and additional carbon rods (forming anodes) are suspended above the furnace. When a powerful electric current is passed through the heated mixture, aluminium is liberated and is deposited on the carbon lining. This pure aluminium is periodically tapped off the bottom of the furnace and cast into ingots ready for further processing.

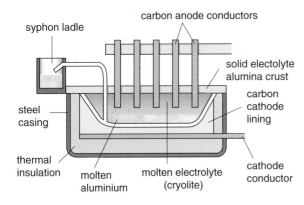

Electrolytic reduction cell

Steel

Steel is produced from iron ore which is also widely found and mined.

The production of steel

To produce steel iron ore must first be processed into iron. The iron ore, limestone and coke are heated in a **blast furnace** using very high temperatures. The limestone is used to remove the impurities from the iron ore.

The iron is added to an **oxygen furnace** where it is converted into molten steel. This molten steel is cast into ingots ready for further processing. In the case of the canning industry, a strip mill will produce large coils of steel as a raw material ready for the making of cans.

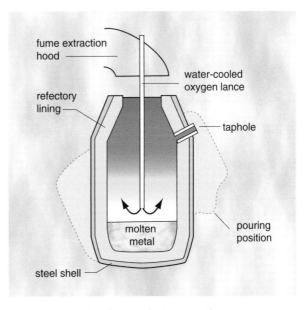

The production of steel using a basic oxygen furnace

One of the main problems with steel is that it can corrode (rust). This is obviously a problem when making food containers, so to prevent it from corroding the surface of the steel is coated with tinplate.

> ### ▪ Hints and tips ▪
>
> Metals can be divided into two groups: ferrous and non-ferrous. There is a quite simple way of identifying them both:
>
> Ferrous metals (i.e. steel) can be picked up with a magnet because they contain iron.
>
> Non-ferrous metals (i.e. aluminium) do not contain any iron and cannot be picked up with a magnet.

> ### ▪ Things to do ▪
>
> 1 At home, use a magnet to identify whether a can is made from a ferrous or non-ferrous metal.
> 2 Make a list of products made from aluminium and a list of those made from steel.

Graphic media

Fine-liner and marker pens

Aim

- To understand a range of graphic media and technical drawing equipment.

The pencil

The 'lead' of a pencil is not in fact made of lead but a graphite composite. This **composite** can be made in varying degrees of hardness and blackness to give hard (H) or soft (B) grade pencils.

Hard pencils

These are pencils that range from grades H to 9H. A hard pencil will have more clay and less graphite content in its lead. This means that it can be sharpened to a fine point which will last a long time. Very hard pencils will mostly be used for technical or more precise drawing where accuracy of line is important.

Soft pencils

These are pencils that range from grades 9B to HB. A soft pencil will have more graphite and less clay content in its lead. This means that the lead will be richer and darker but the point will soften easily. Soft pencils will be used for sketching and shading. A general purpose pencil that can be used for sketching and drawing is the HB.

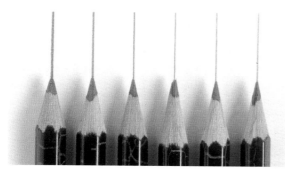

A range of pencils

Fine-liner pens

Fine-liner pens are popular because they have a number of uses – from sketching like a pencil to 'inking-in' technical drawings and from shading small areas to adding notes and lettering to design sheets. Fine-liners will give a good quality line if used properly, but over a period of time the nib can become worn and spread out resulting in a poor quality line.

Marker pens

Marker pens are available in two main types: water-based and spirit-based. Both types are available in a wide range of colours and nib styles including chisel, brush and bullet point. Some more expensive, professional markers are double ended with both a chisel and brush point – there are even markers that contain all three!

Marker pens are widely used to cover larger areas with colour. A good quality, spirit-based marker will leave a flat and solid colour whereas less expensive, water-based types may 'streak' leaving marks of different shades of colour.

Marker pens are excellent for producing **presentation drawings** where a designer will want to communicate the final look of a product.

Drawing-boards

It is important that any designer has a firm, flat surface to work on. A drawing-board should hold your work securely and have a parallel motion (sliding rule) which will aid technical drawing. The standard paper size for GCSE Design and Technology **portfolios** is A3 so your drawing-board must be able to accommodate this size of paper. For the professional designer and architect, there are larger boards available as big as A0 (3 times larger than A3!).

Drawing equipment

There is a wide range of equipment available to aid the drawing of technical graphics.

Computer software packages

Computers are increasingly being used to produce technical graphics or illustrations using computer-aided design.

Photocopier

Black and white photocopiers are extremely useful for making multiple copies of documents. Special features on photocopiers also make it possible to enlarge and reduce your work to the required size. Documents with several pages can also be photocopied back to back, sorted and stapled if necessary. Colour photocopiers are available for copying in full colour, but copies are quite expensive.

Airbrush

Airbrushing can create many interesting and decorative finishes on a variety of materials. Airbrushes produce areas of flat colour or intricate details and are ideal for illustrating surface finishes. It requires great skill and practise to achieve good results. Many professional illustrators use airbrushes.

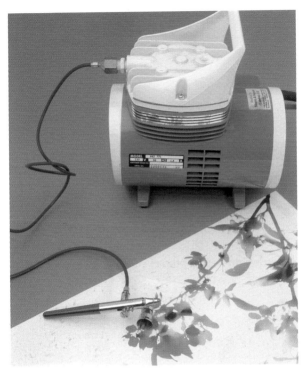

An airbrushed illustration

> ### ▪ Things to do ▪
>
> 1 Use technical drawing exercises to develop your technical graphics.
> 2 Render simple 3D drawings using pencils and marker pens to develop your illustration techniques.

Name	Equipment	Description
Set squares		A **drafting aid** available in 45° or 30/60° used with a drawing-board and parallel motion to produce technical drawings
Plastic rule		A straight edge for drawing lines with a scale (mm) for measuring. A transparent rule is useful for constructing a series of evenly spaced parallel lines, i.e. for cross hatching
Compasses		For drawing circles and arcs. A spring bow compass is ideal for drawing small circles and a pencil compass for larger circles. Adaptors can be fitted to some compasses allowing them to hold fine-liners
Eraser		Used to remove construction lines from technical graphics or correcting pencil-drawn mistakes. Many types are available, but always choose one that does not smudge
Templates		A wide range of shapes and sizes available from simple circle and ellipse templates to standard architectural symbols. Templates provide a quick and effective way of drawing rather than by freehand or technical construction methods
Curves		French curves provide a means of repeating a particular curve without having to construct it technically. Usually available in sets of three. A flexi-curve is a plastic strip with a lead core that can be bent into any desired curve

Drawing equipment

Choosing paper and board

Aim

- To understand the criteria for choosing paper and board for various applications.

Paper

Choice of paper is important in how printed materials look. Choosing the right paper for the job is a combination of personal preference and common sense. The right paper must satisfy:

- the **design** requirements, i.e. surface finish, colour, size and weight

- the demands of the printing process or surface decoration
- economic considerations (price).

Grid papers

Grid papers are available in a range of styles and are useful for generating design ideas when traced through. Apart from the usual squared grid papers used in maths, there are **isometric** and **perspective** grid papers to aid drawing in three dimensions.

> **■ Hints and tips ■**
>
> Gloss-coated inkjet papers have a major advantage over laminated paper because they can be recycled.

Paper	Weight	Description	Uses	Advantages	Cost
Layout	Around 50 gsm	Thin translucent paper with a smooth surface	Outline sketches of proposed page layouts. Sketching and developing ideas	Translucent property allows tracing through onto another sheet. Accepts most drawing media (except paints)	Relatively expensive
Tracing	60/90 gsm	Thin transparent paper with a smooth surface. Pale grey in colour	Same as layout paper. Heavier weight preferred by draughtsmen	Allows tracing through on to another sheet in order to develop design ideas	Heavier weight can be quite expensive
Copier	80 gsm	Lightweight grade of quality paper good	Black and white photocopying and printing from inkjet/laser printers. Smooth finish for colour printing. General use for sketching and writing	Fairly cheap to buy in large quantities. Bright white and available in a range of colours	Inexpensive when bought in bulk
Cartridge	120–150 gsm	Creamy-white paper. Smooth surface with a slight texture	Good general purpose drawing paper. Heavier weights can be used with paints	Completely opaque. Accepts most drawing media	More expensive than copier paper

Drawing papers

Inkjet Paper	Weight	Description	Uses	Advantages	Cost
Coated	80–150 gsm	Bright white, high density, ultra-smooth coated paper	Printing photo quality work with a matte finish, i.e. presentation materials, reports, colour reproductions, etc.	Suitable for 1200 dpi colour inkjet and laser printing. Quick drying. Recyclable	Expensive (usually sold in small packs)
Photo glossy	140–230 gsm	Bright white, professional quality, specially coated high gloss paper	Vivid photo quality with maximum colour reproduction suitable for photo reproductions, graphic artwork and presentation materials	Special coating makes it ideal for digital or scanned images with high resolution. Quick drying. Photo quality. Heavyweight. Two-sided photo gloss paper also available	Expensive (usually sold in small packs)

Inkjet papers

Inkjet papers

Although smooth finished copier paper can be used for black and white printing, there are a number of papers specifically designed for colour printing.

Card and board

There is a wide range of card and boards for a variety of applications. These range from stationery uses such as drawing, writing, photocopying and printing to more creative uses with speciality materials.

Cartonboard

Cartonboards are usually used for retail packaging. These boards must be suitable for high quality, high speed printing and for cutting, creasing and glueing using very high speed **automated** packaging equipment.

Advantages of using cartonboard include:

- total graphic coverage and excellent print quality
- excellent protection in structural packaging nets
- relatively cheap to produce and process
- can be recycled.

■ Things to do ■

1 Tear the board on a cereal or soap powder box and you will see its layered structure quite clearly.

2 Collect samples of different types of paper and board to add to your notes.

Card/board	Weight	Description	Uses	Advantages	Cost
Card	230–750 microns	A thin variety of board, but thicker than paper	A range of uses from printing and drawing to 3D modelling and presentation work	A large range of colours and surface finishes available including bright and fluorescent colours, duo-tones and metallics and corrugated card types	More expensive than paper. Speciality cards are more expensive than simple white or bright colours
Mounting board	1000–1500 microns	Extremely thick board with colour on one side only (white on back)	Mounting work for presentations and displays. Work can be mounted flat or behind a frame mounting	Very high quality, strong and rigid board. Available in a range of colours (wide range of pastel colours)	Expensive

Card and board

Board	Description	Uses	Advantages	Cost
Folding boxboard	Usually has a top surface of bleached virgin pulp, middle layers made from unbleached pulp and a bleached pulp inside layer	Widely used for the majority of food packaging and for all general carton applications	Excellent for scoring, bending or creasing without splitting Excellent printing surface	Inexpensive
Corrugated board	Made from sandwiching a fluted paper layer between two paper liners	Protective packaging for fragile goods. The most commonly used box making material	Excellent impact resistance Has excellent strength for its weight Low cost Recyclable	Inexpensive
White-lined chipboard	Has top layer of bleached wood pulp and a middle and back layer made from waste paper	Food packaging and general carton applications. White-back versions are known as Triplex	Good for high-speed printing of automatically packed cartons	Inexpensive
Solid white board	Made entirely from pure bleached wood pulp	Packaging for frozen foods, ice-cream, pharmaceuticals and cosmetics	Very strong and rigid Excellent printing surface	Expensive
Cast-coated board	A heavier and smoother coating applied to white-lined chipboard and solid-white board	Luxury products requiring expensive looking decorative effects	Very strong and rigid Excellent printing surface Higher gloss finish after varnishing	Expensive
Foil-lined board	Has a laminated foil coating (can be used on all of the above boards)	Cosmetic cartons, pre-packed food packages	Foil available in matt or gloss finish and in silver or gold colours Very strong visual impact Foil provides an excellent barrier against moisture	Expensive

Common cartonboards

Choosing plastics

Aims

- To understand why plastics are used for packaging.
- To understand the uses of different plastics for packaging.
- To understand the properties and uses of expanded polystyrene.

Plastics in packaging

Plastics are widely used in packaging because they are:

- versatile
- lightweight
- low cost
- energy saving
- tough and durable
- recycleable.

Plastics can be identified by a coding system usually stamped on to the base of the package or on the label. This is an internationally recognized system that enables plastics to be easily identified for recycling.

For example, a shampoo bottle stamped with the opposite identification mark would be made out of a high density polyethylene plastic and could therefore be sorted visually.

HDPE

Each plastic has its own useful properties that make it suitable for use in different areas of packaging (see the table 'Common thermoplastics in packaging' opposite).

Plastics versus glass

Glass, once one of the most common materials used for bottling, has increasingly been replaced by the use of plastics.

In the obvious case of fizzy drinks bottles, glass is a suitable material, but plastics have proved more successful because they are cheaper, lighter, durable and do not smash if dropped.

In the case of the ketchup bottle, the development of a new type of plastic made up of several layers makes it possible to have a squeezable bottle type – unthinkable with glass.

Expanded polystyrene

When we look at a throw-away society we often look towards the USA as a prime example. This is the home of fast food with hundreds of fast-food chains available including the major companies of McDonald's and Burger King.

Polystyrene is used in fast-food packaging because it is:

- hygienic
- strong, yet lightweight
- efficient
- economical
- convenient.

Hygienic

Tests in the USA (Polystyrene Packaging Council) into the use of disposable polystyrene food service ware such as cups and plates have found that they are more sanitary than reusable service ware. In other words, germs and bacteria are simply thrown away with the rubbish instead of multiplying in a chipped coffee mug.

Strong, yet lightweight

Polystyrene protects against moisture and keeps its strength even after long periods of time. Containers and lids close tightly and prevent any leakage of the contents. It can be moulded into a variety of structural packages which compliment its excellent cushioning properties in protecting the contents of the package.

Efficient

Polystyrene provides excellent insulation. Therefore, hot food can be kept warm for longer periods of time. It also means that the package does not become so hot that it cannot be held in the hand.

Economical

Polystyrene food service products are generally cheaper to buy than disposable paper products and much cheaper than reusable service ware (i.e. china). This is because only about five per cent of the foam package is actually plastic – the rest is simply air!

Convenient

This is arguably the major reason for the use of polystyrene in fast-food packaging. With today's busy lifestyles people want food to be available instantly, and polystyrene is an economical way of serving people with their fast food.

Thermoplastic	ID code	Properties	Applications
PET Polyethylene terephthalate	1 PET	Excellent barrier against atmospheric gases and does not allow gas to escape Does not flavour the food or drink contained in it Sparkling 'crystal clear' appearance Very tough Light – low density	Carbonated (fizzy drinks bottles) Packaging for highly flavoured food Microwaveable food trays
HDPE High density polyethylene	2 HDPE	Highly resistant to chemicals Good barrier to water Tough and hard wearing Decorative when coloured Light and floats on water Rigid	Unbreakable bottles (for washing-up liquid, detergents, cosmetics, toiletries, etc.) Very thin packaging sheets
PVC Polyvinyl chloride	3 PVC	Weather resistant – does not rot Chemical resistant – does not corrode Protects products from moisture and gases while holding in preserving gases Strong, abrasive, resistant and tough Can be made either rigid or flexible	Packaging for toiletries, pharmaceutical products, food and confectionery, water and fruit juices
LDPE Low density polyethylene	4 LDPE	Good resistance to chemicals Good barrier to water, but not to gases Tough and hard wearing Decorative when coloured Very light and floats on water Very flexible	Stretch wrapping (cling film) Milk carton coatings
PP Polypropylene	5 PP	Lightweight Rigid Excellent chemical resistance Versatile – can be made stiffer than polythene or very flexible Low moisture absorption Good impact resistance	Food packaging – yoghurt and margarine pots, sweet and snack wrappers
PS Polystyrene	6 PS	Rigid polystyrene: • Transparent (clear) • Rigid (can be brittle) • Lightweight • Low water absorption Expanded polystyrene (foam): • Excellent impact resistance • Very good heat insulator • Durable • Lightweight • Low water absorption	Food packaging, (i.e. yoghurt pots), CD cases, jewel cases, audio cassette cases, take-away food packaging, egg cartons, fruit, vegetable and meat trays, cups, etc. Packing for electrical and fragile products

Common thermoplastics in packaging

Other uses

Expanded polystyrene has another important use in the protective packaging of many products. It comes in two main forms:

• loose-fill 'peanuts'
• shape-moulded packaging.

Loose fill 'peanuts' allow various sized products such as stationary to be transported in the same box without them being damaged. Shape moulded packaging fits snugly around delicate products and the manufacturer's advertising is clearly printed on the cardboard box into which it fits.

Shop shelves are full of electrical products that are protected by expanded polystyrene, but did you know that a Formula 1 racing car can also be packaged for export?

▪ Things to do ▪

1 Compile a chart of household plastic bottles and the type of plastics they are made of. Why do you think they are made of these particular plastics?

2 Expanded polystyrene is a very useful packaging material for fast food, but what are its disadvantages?

Choosing wood

Aims

- To understand the qualities of manufactured boards, softwoods and hardwoods in the production of high quality products.
- To understand the need for a suitable mould when vacuum forming.

Product modelling

The most important role wood has to play in a graphic product is in the manufacture of a three-dimensional (3D) **prototype**. MDF is an ideal material for producing a high quality product because it:

- can be shaped easily
- has an excellent surface finish.

When designing a product such as a mobile phone, computer mouse or FM radio, MDF can be formed into smooth streamlined shapes essential for modern looking products. MDF is available in sheets usually 9–24 mm thick, but can be glued together (laminated) using PVA to achieve greater thicknesses.

The MDF block can be cut to a rough shape by marking on the plan and side profiles and cut on a bandsaw. Your teacher will have to cut the shape out using the bandsaw as it is illegal for you to do so.

When the rough shape is cut out, it is possible to shape the MDF using tools such as surforms to achieve the desired product styling.

Cutting MDF block to a rough shape

Shaping MDF using a surform

The prototype model can then be sanded extremely smooth using various grades of glasspaper.

■ Hints and tips ■

At this stage the MDF has to be sealed using a sanding sealer so that any surface finish applied will not soak into the fibres. Once dry, lightly sand the surface to achieve a smooth and level surface.

A quality finish is achieved by using an acrylic spray primer and sanding it back gently using wet and dry paper. Good quality acrylic car paints are available in a wide range of colours to apply a professional looking top coat.

Achieving a professional looking finish using acrylic car paints

Some products will be circular in shape and will therefore need to be turned using a wood lathe. Softwoods such as pine can be used for such **product models** because they are available in square sections. Once mounted on a wood lathe, it is possible to create some very interesting shapes indeed.

The classification and selection of materials and components

In the photo below, the student first turned the main shape on a wood lathe and then carved the detail of the sweeping speaker grill using a gouge.

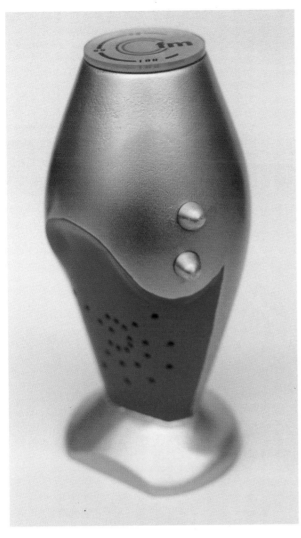

A student's model of an FM radio

Hardwoods are not usually used for product modelling as they are expensive and difficult to shape. They are usually used for products where their excellent surface finish is desirable, for example furniture.

Vacuum forming moulds

MDF and pine can also be used for the production of moulds for vacuum forming, usually for the purpose of blister packaging.

In much the same way as producing a product model, a mould can be cut and shaped to create interesting shapes. MDF is the most suitable wood because it has no grain. This means that the mould will not leave an imprint on the vacuum-formed plastic shape.

The suitable mould needs careful consideration when designing and making. The mould must:

- be very smooth
- have slightly angled sides (usually 5°)
- have rounded or 'radiused' corners and edges.

This will ensure that the mould can be easily removed once vacuum formed.

Interior and architectural modelling

A variety of woods can be used for the production of an interior or architectural model. Manufactured boards are useful when creating walls or partitions, whereas the use of hardwoods can give high quality details.

In the photo below, the student has constructed an architectural model using birch-veneered plywood to give a high quality finish.

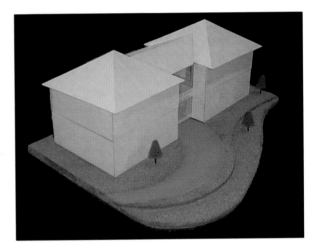

A student's architectural model

■ Hints and tips ■

Cheaper manufactured boards such as hardboard could also be used and either painted or covered in brick-effect paper to give an alternative finish.

■ Things to do ■

1 Collect examples of each enhancement technique described in this section.

2 Discuss the reasons why a manufacturer may use expensive enhancement techniques on their products.

Choosing metals

Aims

- To understand the benefits of metals for preserving food.
- To understand the advantages of using metals in packaging.

A range of metal product packaging

Aluminium or steel?

Metal packaging is mainly used in closures (screw caps, etc.), food cans and beverage cans. Closures and food cans are usually made from steel and beverage cans are divided equally between aluminium and steel (quantities and percentages vary on a yearly basis).

Steel is commonly used for packaging drinks, aerosols, processed and powdered foods, chocolate and biscuits, paints, adhesives and chemicals, health and beauty products, giftware and closures.

Aluminium is commonly used for packaging drinks, aerosols, health and beauty products, tamper-proof closures and screw caps.

There are several advantages in using both aluminium and steel for packaging:

- Security – sealed cans cannot be tampered with.
- Packages can be made in a variety of sizes and shapes including cylindrical, rectangular and hexagonal.
- Packages can be **embossed** to provide surface detail.
- Metal can be directly printed on to or a paper label added.
- The packaging itself offers **point-of-sale** display with all over decoration (product recognition).

Metal appeal

Metal packaging suggests quality, and manufacturers will often use it for special promotions. For example, during the Christmas period many brands of chocolate and biscuits will be available in large and highly decorative metal tins. Some biscuit tins dating from the nineteenth century have even become collectors' items.

The production of metal cans

Two types of cans widely used are:

- the three-piece welded can (tinplate) used for processed food, for example baked beans
- the two-piece drawn and wall ironed can (either tinplate or aluminium) for beverages, for example soft or alcoholic drinks.

Cold forming an aluminium drinks can

The forming of the two-piece drinks can involves two main stages:

1 Drawing – an aluminium disc is pressed under high pressure into a die to form a shallow cup shape.

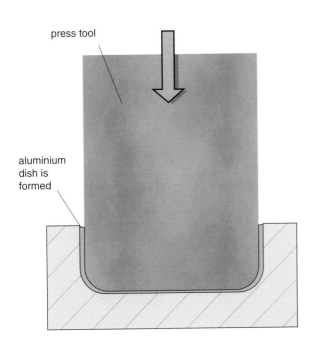

2 Ironing – the aluminium cup shape is drawn down into a deeper die to form the base profile. At the same time an ironing ring thins out the walls of the drinks can.

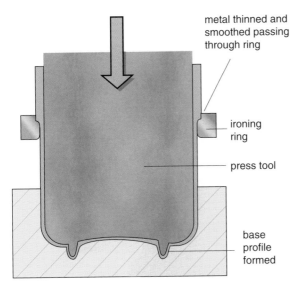

Advantages of the aluminium drinks can

- Paper thin walls (0.1 mm) save materials and energy
- Aluminium with internal lacquered coating does not react with the contents
- Able to withstand the high internal pressures of a fizzy drink
- Lightweight
- Relatively inexpensive to produce in quantity
- Design can be printed directly on to the outside of the can once formed.

An aluminium drinks can

As materials technology has developed, can manufacturers have been able to reduce the amount of materials used in the manufacture of a single can. By calculating the stresses acting upon the can the thickness of the material has been reduced and a tapered neck saves even more material without lessening its structural ability.

> ### ∎ Things to do ∎
>
> 1 Research pictures of old metal packaging and see how they compare with modern metal packaging.
>
> 2 Test the strength of a drinks can by (a) crushing the sides using one hand and (b) putting all your weight on top of it (use two friends either side of you to help you balance).

Choosing glass

Aims

- To understand the properties of glass as a packaging material.
- To understand the strong brand identity created by Coca-Cola using the glass 'hobbleskirt' bottle.

Colourant	Glass colour(s)
Iron	Green, brown, blue
Manganese	Purple
Chromium	Green, yellow, pink
Vanadium	Green, blue, grey
Copper	Blue, green, red
Carbon and sulphur	Amber, brown

Glass has many properties that make it ideal for use as a packaging material:

- It is relatively cheap when mass produced.
- It is resistant to mechanical shock.
- It has excellent product visibility.
- It offers excellent protection against contamination.
- Its contents can be preserved through high-temperature processing.
- The lid provides an air tight seal
- It can be re-used and recycled.

Colour

In its purest form, glass has a greenish tint. By adding chemicals in varying quantities to the raw mixture of sand, soda and limestone different colour glass can be produced. This is useful when designing glass containers for a specific product. For example, beer is usually packaged in green or brown glass to protect it from direct sunlight.

Glass versus plastics

Glass has one major advantage over the use of plastics – it looks expensive and therefore has the image of sophistication. For example, to use a plastic bottle for an expensive vintage champagne would be inappropriate. Glass can be formed into interesting shapes that resemble expensive crystal glasses and give the product a look of quality.

Could an expensive product like champagne be packaged in any material other than glass?

The Coca-Cola 'hobbleskirt' bottle

The use of glass in packaging is perhaps best illustrated by the Coca-Cola bottle. Who could mistake the shape of the Coca-Cola bottle? Even in silhouette, this classic bottle design is easily recognizable. Modern plastic (PET) Coca-Cola bottles have been designed to resemble the glass bottle because of its strong brand identity.

An early Coca-Cola bottle

The Coca-Cola 'hobbleskirt' bottle has become an icon of the twentieth century

John Pemberton, who invented Coca-Cola in 1886, originally used plain bottles with paper labels to sell the Coca-Cola syrup to shops or 'soda fountains'. At fountains the syrup was mixed with plain water and served to customers. Later, carbonated water was used and bottled so that people could enjoy the soft drink away from the soda fountains.

The early bottles were not marked with the Coca-Cola trademark. Instead, they simply used existing bottles from various manufacturers. By the end of the nineteenth century, a variety of bottles had been introduced, some of which had the Coca-Cola trademark blown into the glass during the forming process.

However, these early bottles did not give Coca-Cola a strong brand image and the need for a standardized bottle was finally considered. A Coca-Cola company executive at the time stated: 'We need a bottle which a person can recognize as a Coca-Cola bottle when he feels it in the dark'. Bottle manufacturers were invited to submit designs for the new bottle and the winner was chosen at the annual 1916 Coca-Cola Bottlers Convention.

The winning design was based upon the shape of a cocoa pod with an exaggerated bulge around the middle. The original prototypes had to be modified in order to fit automatic bottling equipment by slimming the bottle to its now classic contour shape.

By 1920, the new standardized bottle – called a hobbleskirt bottle because its shape resembled a dress fashion of the day – was in widespread use throughout the United States.

> ### ▪ Things to do ▪
>
> 1 Make a list of all the products you can think of that use glass packaging. Explain why the manufacturer has used glass rather than any other material.
>
> 2 Find out if there are any glass recycling schemes in your local area.
>
> 3 Design a new glass bottle for the packaging of a new soft drink

Components

Aims

- To understand the use of components for adding surface detail and texture.
- To understand the use of components for securing and fixing graphic media.

*This prototype of a handheld computer games console uses all three **aesthetic** components*

Aesthetic components

There is a range of components available for adding surface detail and texture to block models in order to make them look more realistic.

Component	Example	Applications	Advantages	Disadvantages
Modellers' raised plastic lettering		Professional looking raised numbers or lettering for prototype models	Gives the appearance of low relief moulding on models as produced by the injection moulding process	Expensive Tricky to remove from **sprue** and apply to surface of model
Self-adhesive paper labels		For adding surface texture or decoration to models, i.e. representing grip texture	Available in a range of different styles and colours Cheap	Tricky to apply in straight lines to emulate grip texture on models
Dry-transfer letters		Rub-down lettering for adding flat numbers and lettering to the surface of block models or paper and card models	Relatively easy to apply Available in a wide range of fonts, colours and sizes and in architectural symbols, textures, vehicles and people	Expensive – the use of computer-generated text may be an alternative

Aesthetic components

Component	Example	Applications	Hints and tips
Paper fasteners		Creating pivots for card mechanisms or ergonome models	Do not tighten Allow all moving parts to move freely
Paper clips		Temporarily attaching pieces of paper and card, i.e. securing tracing paper over a drawing when copying	Avoid marking paper with one of points or crinkling paper
Drawing pins		Attaching presentation work to display boards	Press firmly into material without bending the drawing pin
Mapping pins		Attaching presentation work to display boards Indicating the positions of important information, i.e. on a map or diagram Securing the jointing of foamboard walls on a model while glueing	Use only on soft display board material (will bend in hard material) Use various colours to indicate different features

Functional components

Method	Picture	Description	Applications	Application in schools
Saddle-wire stitched		The simplest method of binding by stapling the pages together through the fold	Documents for presentation	Use a long-arm stapler to enable you to staple the document through the fold
Side-wire stitched		Staples are passed through the side of the document close to the spine. Used when the document is too thick for saddle-wire stitching	Documents for presentation	Use an ordinary stapler to staple the document along its side
Perfect binding		Pages are held together and fixed to the cover by means of a flexible adhesive. This method produces a higher quality presentation and the spine can also be printed on	High quality documents for presentation, magazines, less expensive books	Not possible in school. Take document to a commercial printer for binding
Hard-bound or case-bound		Usually combines sewing and glueing to create the most durable method of commercial binding. Stiff board is used on the cover and back to protect the pages	Books	Not possible in school. Take document to a commercial printer for binding
Spiral or comb-binding		Pages are punched through with a series of holes along the spine. A spiralling steel or plastic band is inserted through the holes to hold the sheets together	Documents for presentation	Use a comb binding machine

Binding methods

Functional components

Many graphic products require the use of components for securing and fixing graphic media such as paper and card. They include:

- paper fasteners
- paper clips
- drawing pins
- mapping pins.

These are all described opposite and are inexpensive to buy.

Binding methods

There are five main methods of binding a brochure, magazine or book. They are:

- saddle-wire stitched
- side-wire stitched
- perfect binding
- hard-bound or case-bound
- spiral or comb-binding

Some of these will be suitable when producing printed materials in your projects. These methods are all described in the table above.

▪ Things to do ▪

1 Make a simple card linkage mechanism using paper fasteners as pivots.

2 Use a binding method to present a product analysis report.

Glass in packaging

Glass is one of the earliest materials used for containing food and drink and continues to be used for a range of packaging uses.

The ingredients of the most common commercial glasses are sand, limestone and soda. These ingredients are heated at a temperature of around 1500°C and react to form a liquid. This liquid can then be moulded into shape and allowed to cool so it forms a hard, inert and transparent material. However, glass can be made in a variety of ways containing different chemicals to produce glasses with different properties and colours for varying uses.

Food storage

Glass containers have been used since Roman times as a means of storing food. It was not, however, until the nineteenth century that glass was used to help preserve foods. It was discovered that certain foodstuffs such as fruit, meat, fish and vegetables when heated at high temperatures and sealed in glass jars could be preserved for long periods of time. Glass was an ideal material as air could not penetrate through it and spoil the food.

Bottles for soft drinks

Until the beginning of the seventeenth century, nearly all bottles were made of earthenware, metal, wood or leather. Early stoppers were made out of wax, later replaced by cork. The earthenware bottles used by the early mineral water manufacturers were unsatisfactory as the gas could escape at high pressures. In 1814 the first egg-shaped glass bottle was patented for bottles of artificial mineral water or 'pop' which had a much greater resistance to internal pressures. Later, the now famous Codd bottle was introduced. It contained a glass marble that was kept pressed against a rubber ring in the neck of the bottle by the internal gas pressure resulting in an air tight seal. Nowadays, most glass bottles either have a crown cap or screw cap for convenience.

Early egg-shaped and Codd bottles

Milk bottles

Until World War I, milk was sold from churns pushed around the streets in hand carts. This could be extremely unhygienic, so shortly after the war milk was pasteurized (sterilized to kill bacteria by heating) and sold in sealed glass bottles. Milk is still available in glass bottles with aluminium tops, but this is becoming increasingly uncommon due to the availability and cost of large plastic milk bottles in supermarkets.

Manufacturing glass containers

Early glass containers were blown by hand by craftsmen, but they were usually thick and heavy because the blower could not control the glass distribution. Modern manufacture uses automatic processes for **mass production**.

Raw materials are automatically mixed and fed into the furnace where they are heated and fused at approximately 1500°C

Molten glass is fed into a machine where it is automatically blown. Bottles are made in two stages. First, a **parison** shape is blown. This is transferred to a second mould in which the bottle is blown to its final form

Bottles are inspected and despatched for filling, capping and labelling

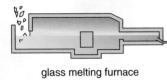

glass melting furnace

tunnel called a lehr in which bottles are reheated and gradually cooled to prevent stresses developing

Making glass containers by automatic process

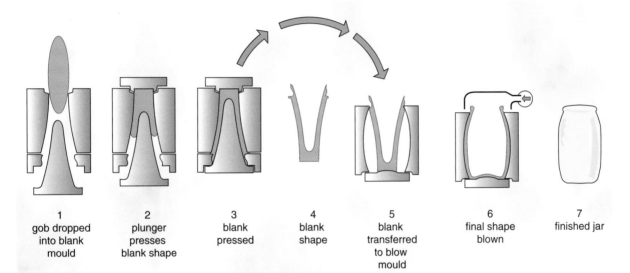

1	2	3	4	5	6	7
gob dropped into blank mould	plunger presses blank shape	blank pressed	blank shape	blank transferred to blow mould	final shape blown	finished jar

The automatic press and blow process

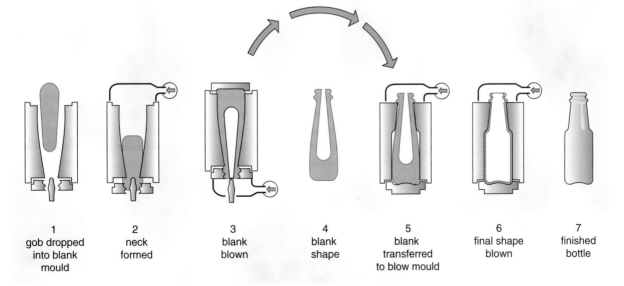

1	2	3	4	5	6	7
gob dropped into blank mould	neck formed	blank blown	blank shape	blank transferred to blow mould	final shape blown	finished bottle

The automatic blow and blow process

1 Packaging is an important part of a product.

a Complete the table below by:

i naming three more packaging materials

ii providing a specific example for the use of the named material. **(6 marks)**

Packaging material	Example of packaging
Paper and boards	Easter egg boxes

b Corrugated card and expanded polystyrene are often used to package electrical products. For each of these materials describe its application and characteristics that makes it useful for this purpose. **(6 marks)**

c Solid board is another material used for packaging electrical products. Give two advantages and one disadvantage of its use instead of corrugated card. **(3 marks)**

2 The diagram below shows a shop window display for the promotion of a new video release.

a Two points of specification are:

- the Astro-man figure should be life sized

- the whole display should be lightweight for easy installation/transportation.

Give two more points of specification which would be in the specification for this display. **(2 marks)**

b Name the type of material suitable for:

i printing the computer graphic on

ii making the Astro-man figure

iii making the Astro-man lettering. **(3 marks)**

c Give one property associated with one of the materials you have named in (b) and explain how this property makes it suitable for this application. **(4 marks)**

3 A plastic drinks bottle is shown to the right.

a Medium density fibreboard (MDF) or polystyrene block may have been used for making a prototype of this bottle.

i Give two advantages of MDF over polystyrene for this purpose. **(2 marks)**

ii Give one advantage of polystyrene over MDF for this purpose. **(1 mark)**

b The bottle is made from PET. Describe three properties of PET that relate to its intended use. **(6 marks)**

Section B:
Preparing, processing and finishing materials

The Rocky 4 Mars Pathfinder Rover prototype goes through its paces at the jet propulsion laboratory. The Pathfinder Mission to Mars in 1997 was a much publicized success.

Corrugation

Aims

- To show that corrugation is a strengthening technique.
- To illustrate the various forms of corrugated board and their properties.

The basic principle

Corrugation is a strengthening technique where a sheet of material is shaped into alternate ridges and grooves. For example, a flat sheet of card is very easy to bend, but when bent into a series of 'V' shapes it becomes rigid and harder to bend.

Strengthening paper through corrugation

Corrugated board

Corrugated board is made from layers of paper. The top and bottom surfaces are called liners and the corrugated internal layer is known as fluting. There are three main combinations of corrugated board: single, double and triple wall board. They each have different properties.

Making corrugated board

There are three main stages in the manufacture of corrugated board:

1 Processing of wood to produce semi-chemical paper, then forming the paper into flutes
2 Processing of wood to produce Kraft papers for liners
3 Bonding the fluting and liners together.

Fluting

Semi-chemical paper is used for making the fluting in corrugated board. Birch hardwood is usually used in Europe as it is quick growing and easy to farm. The birch trees are fed through machines to produce wood chips and the fibres are separated using both mechanical and chemical methods. The fibres are then made into paper using a papermaking machine.

Semi-chemical fluting requires heat, moisture and pressure in the corrugator roll nip to bend and move the fibres into the flute shape (see diagram).

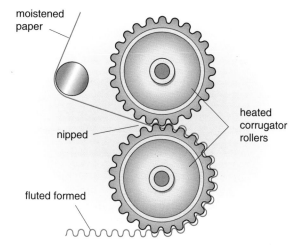

The flute forming process

Board	Characteristics	Applications
Single wall board	Coarse – high stacking strength from large flute profile Fine – good printing surface; uses less fluting than coarse Extra fine – very good printing surface from closely pitched flutes	Containers for transporting and storing goods, point-of-sale displays
Double wall board	Very good stacking strength from rigid board Very good resistance to shock and puncture Good printing surface	Containers for transporting heavy weight goods and goods needing maximum protection
Triple wall board	Very good stacking strength from very rigid board Very good resistance to tear and puncture Long life (up to 10 years' storage capability) Multi-trip container (reusable)	Transportation of bulk food, chemicals, heavy engineering components, automotive parts, electrical and electronic equipment

Corrugated boards and their characteristics

The paper must have an 8–9 per cent moisture content as it is fed through heated corrugator rolls at 180°C. The nip pressure at these rollers then forms the paper into the distinctive flute profile. The flute profile can be altered to produce various flute types and sizes.

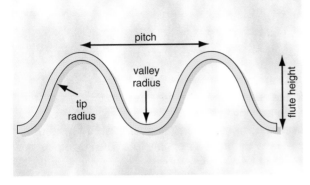

The flute profile

Flute		
Coarse	3.2–4.8 mm	105–145
Fine	2.1–3.0 mm	150–185
Extra fine	1.0–1.8 mm	290–320

Standard flute sizes

Kraftliners

The liners are made from **Kraft papers** which have a smooth outer printing surface and a slightly coarser inner glueing surface. Kraftliner is made from at least 80 per cent new softwood pulp and processed into paper at the paper mill.

The wood pulp is sometimes dyed brown to give its distinctive colour or made from bleached fibres to give a bright white appearance. Alternatively, a layer of bleached white fibres can be added to a layer of brown fibres in the papermaking machine or a clay/latex coating can be added.

Recycled paper

The use of recycled paper for making both the liner and fluting is now very common. Recycled paper is made from waste paper such as old corrugated cases, newsprint and magazines which are mixed with water, broken up in a pulper, cleaned and processed. They can be made into new single-ply or multi-ply papers.

Bonding

The liners are bonded with the internal fluting by the use of a starch adhesive. Starch is produced from maize, wheat, or potatoes. Chemicals are added to reduce the point at which it gels (or becomes a solid adhesive).

The flute tips pick up the starch from an applicator roll and the starch absorbs into the flutes. The liner must be warm and damp when the liner and flute tips are pressed together firmly. The starch then gels and the moisture moves out of the starch and into the papers to form a secure bond.

■ Things to do ■

1 Try making your own corrugated board from two pieces of paper as liners and a concertinaed piece of paper to act as the internal fluting.
2 Look out for the use of single, double and triple walled board in corrugated cases.

■ Hints and tips ■

- Coarser flutes have greater stacking strength.
- Finer flutes produce a better printing surface.
- Double wall board combines both advantages.
- Triple wall board has a very high stacking strength and is long-lasting.

Type	Ply	Printing surface	Cost
Brown Kraft (unbleached)	Single ply or multi-ply at least 80% new fibre	Fairly good	Expensive
White Kraft (bleached)	Single ply or multi-ply strong new fibre	Very good	Very expensive
White-top Kraft (bleached top ply)	Multi-ply strong new fibre	Very good	Expensive, but cheaper than bleached Kraft of same weight
Coated Kraft	Clay/latex coating on bleached Kraft and white-top Kraft papers	Excellent	Very expensive

Kraftliner types

Laminating

Aims

- To describe the properties of laminating in wood, plastics and paper.
- To explain the characteristics of materials used in laminates for packaging.

The basic principle

Lamination is both a strengthening and decorative technique. This process is used in a range of products and in different materials.

Laminating wood

Timber is a useful structural material which has two major disadvantages:

- It is not uniform in strength due to knots and other defects.
- It is not always available in large sizes and, when available, is costly.

Defects can be overcome by laminating – taking a number of thin pieces (veneers) and glueing them together to form beams or sheets (plywood).

In the case of plywood the layers are placed with the grain running in alternate directions to give strength in both directions. By bending laminated layers of ply around a former it is possible to create curved shapes.

Laminating plastics

The lamination of thermoplastics such as acrylic can add strength (acrylic is very brittle) or more importantly add a decorative effect. By laminating several pieces of acrylic sheet together using Tensol cement, it is possible to create a multi-coloured 'liquorice allsort' effect. The block can then be finished and polished to a high sheen.

Laminating paper and card

Laminating paper or thin card can enhance the appearance and add protection to printed materials. The plastic surface finish makes the card more durable, such as bus passes where the card is used regularly, and also gives it a wipe-clean surface for use on menus in restaurants.

Laminated plywood is used in skateboard decks to form the curved shape and to add strength

Laminated paintings decorate Yemeni suitcases

Packaging laminates

Laminates are used in packaging to great effect. Plastic films or coatings are added to materials in order to produce different characteristics.

Plastic coatings are added to cartons to give extra protection. A layer of polythene is added which protects the food or liquid contents from contamination by air or moisture. For an even longer life, a layer of aluminium foil is added.

Aluminium foil

Aluminium foil is an important part of packaging laminates. It is used for a variety of products from chocolate wrappers, tea, coffee and biscuits to pharmaceuticals and healthcare products.

Thin, clear plastic foils are often used in packaging laminates to:

- sandwich and protect a printed surface
- provide an attractive surface finish

- provide an excellent barrier against moisture
- add durability and strength
- enable the product to be heat sealed for hygiene, i.e. air-tight seal to prevent contamination and keep product fresh.

For example, some coffee packaging (filter coffee) uses laminates of polyester films and aluminium foil. The print is protected by printing it on the underside of the film, so it is sandwiched between the film and the foil. The laminate can also be vacuum moulded into a block shape for easy shelf stacking.

Packaging using a foil laminate

Paper, foil and polythene laminates

Paper, foil and polythene laminates combine the characteristics of all three materials into a single package. In general:

- foil provides a barrier to moisture
- polythene can be heat sealed
- paper provides excellent print quality.

> ■ **Things to do** ■
>
> 1 Examine curved pieces of wood to determine whether they have been formed using laminated plywood.
>
> 2 Make a laminated key fob shape from small pieces of sheet acrylic of different colours.
>
> 3 Tear a laminated package to see the cross-section of the laminate.

Printing processes 1

Aims

- To understand the lithographic, letterpress and gravure printing processes.

There are four main printing processes used in commercial printing:

- **lithography**
- **letterpress**
- **gravure**
- **screen printing** (this is looked at on pages 36–37).

Each has its own characteristics that make it suitable for different applications.

Lithography

Lithography is the most widely used process in commercial printing because it is economical, versatile and capable of printing high quality images on a wide variety of papers.

The basic principle

The lithographic process literally means stone writing. The basic principle is that water and grease do not mix. Originally, a grease crayon was used to draw the image on a slab of limestone and then dampened with a water solution. The greasy image rejects the solution and the surrounding area accepts the solution. When the ink is applied to the stone, the image area accepts the ink and the surrounding area rejects the ink. The image is then transferred with a press from the stone to the paper.

Modern offset litho

In modern offset litho (lithography), the flat stone has been replaced by three cylinders: the plate, blanket and impression cylinders. In operation, the aluminium printing plate is dampened with a water solution that the image rejects and the surrounding area accepts.

When the plate is inked, the image area accepts the ink. The inked image is then transferred from the plate cylinder to the blanket cylinder which 'offsets' or prints on to the paper as it passes between the blanket and impression cylinders.

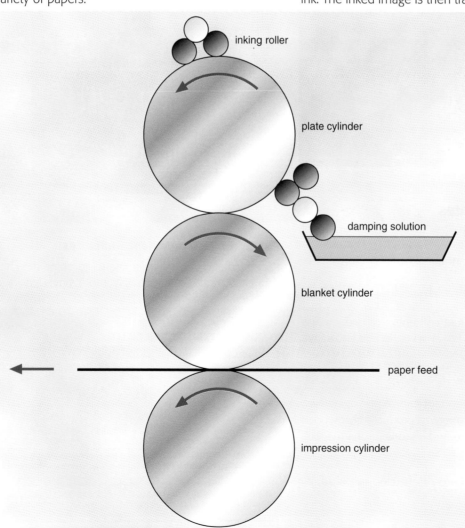

inking roller

plate cylinder

damping solution

blanket cylinder

paper feed

impression cylinder

The offset litho printing process

Letterpress

Letterpress is the oldest printing process of the four, but has been widely replaced by offset litho. Letterpress produces higher quality printed text due to the dense ink used, whereas lithography uses a diluted ink.

The basic principle

Letterpress is a relief printing process which means that the image to be printed is raised above the non-printing background. A dense ink is applied to the raised image and transferred with a press from the printing plate to the paper.

Modern rotary letterpress

The most common form of letterpress in modern commercial printing is the rotary letterpress. Here the printing plate is made from a flexible metal or plastic and clamped to a cylinder. The plate cylinder revolves against ink rollers and in turn makes an impression on the paper that is continuously fed between the plate and impression cylinders.

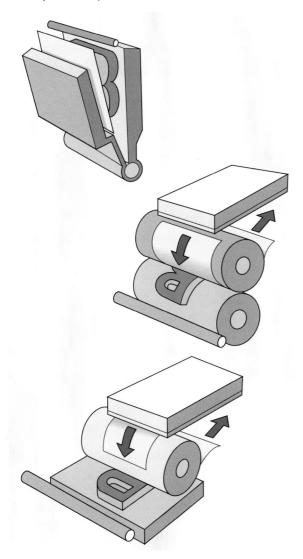

The letterpress process

Gravure

Gravure is used to produce high quality photographic images because of its excellent reproduction of fine detail. Its main disadvantage, however, is that it is a high cost process mainly due to the expense of making the original printing plate.

The basic principle

Gravure is opposite to letterpress in the fact that the printing image is recessed or lower than the non-printing surface. The image is engraved into a copper printing plate creating cells which are filled with a thin, spirit-based ink. The paper is pressed into the ink-filled cells to produce the printed image.

Modern web-fed gravure

Most modern gravure printing is done with web-fed machines which use large reels of paper. The cells are filled with liquid ink and a blade is pulled across the cylinder to remove any excess. As the paper is fed continuously through the press by a rubber covered cylinder, it is pressed into the cells to pick up drops of ink to form the final printed image. The spirit-based ink dries through evaporation immediately after printing.

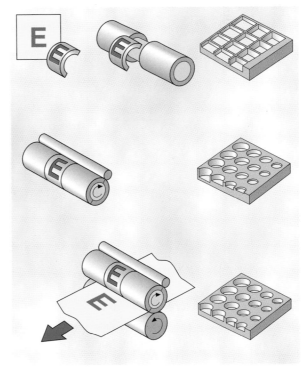

The gravure process

▪ Things to do ▪

1 What does the term 'offset' mean in the offset lithographic process.
2 What is the difference between web-fed and sheet-fed printing processes?.

Printing processes 2

Aims

- To understand the screen printing and flexographic printing processes.
- To understand the advantages and disadvantages of printing processes and the correct printing process for specific materials.

Screen printing

Screen printing is an extremely versatile printing process because it can be used on virtually any type of material.

The basic principle

A stencil is supported on a screen, originally made of silk but now often a synthetic fibre, and stretched tightly over a frame. A thick ink is spread across the screen using a rubber squeegee forcing the ink through the screen and the stencil's printing area on to the paper. The non-printing area of the stencil stops the ink from passing through it and prevents the background from being printed.

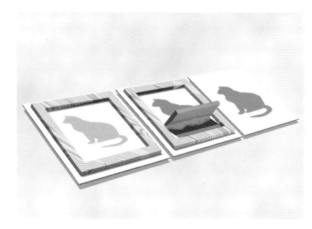

The screen printing process

Manual carousel screen printing

In a manual carousel screen printer the paper (or other material) is held flat on the printing bed by the use of vacuum suction through the bed. Several screens are held on a carousel that can be rotated. The first screen is lowered over the paper and the first coloured ink is applied using a rubber squeegee through the screen and stencil. The first screen is then raised and the carousel rotated on to the next screen. The next screen contains a stencil for a different part of the image requiring another colour. The next colour is printed and the carousel is rotated once more.

By rotating the carousel several times with screens containing different portions of the image and by using several coloured inks, the completed full colour image is built up.

Manual carousel screen printing

Other printing processes

In addition to the four main printing processes, there are a number of other processes in operation for specific uses. For example, when printing directly on to glass, ceramic decoration is used. This is basically a screen printing process using enamels (instead of inks) that are baked at very high temperatures. The most widely used of the other printing processes is flexography.

Flexography

Flexography is similar to letterpress in using a relief plate, but as the name suggests, it uses a flexible plastic or rubber printing plate instead.

Flexography is used mainly for packaging where materials other than paper are used such as PVC for shrink sleeves or foil and foil laminates. It can, however, be used to print any material that will pass through the printing press. Its major application is in the printing of local and national newspapers and less expensive magazines because of high printing speeds and the quick make up of printing plates. Its speed and cheapness have also made it ideal for printing paperback books which, in turn, have enabled paperbacks to become widespread and be sold relatively cheaply.

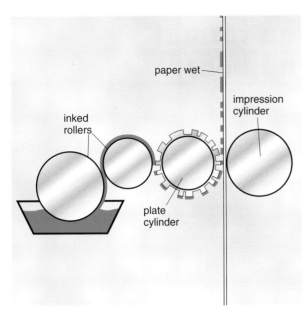

inked rollers

paper wet

impression cylinder

plate cylinder

The flexographic printing process

Each printing process has its advantages and disadvantages that make it suitable for various printing applications.

■ Things to do ■

1 Find out more information on ceramic decoration for printing directly on to glass and dry-offset for printing directly on to metals.

2 Make a set of stencils, cut from card, to simulate the screen printing of an image using three colours.

Material	Commercial printing process
Plastics	On to paper labels – lithographic, gravure, letterpress, screen printing On to shrink sleeves – as above On to stretch labels – as above Directly on to plastic – heat transfer labels, gravure, screen printing, dry-offset
PVC shrink sleeves	Gravure – reverse printed Flexographic – reverse printed
Glass	On to paper labels – lithographic, gravure, letterpress, screen printing On to polypropylene (PP) labels – as above Directly on to glass – ceramic decoration
Metals	On to paper labels – lithographic, gravure, letterpress, screen printing Directly on to metal – dry-offset print, reprotherm (transfer system for full colour photographic image)
Solid board	Lithographic Flexographic Screen printing Gravure
Foil/film laminates	Flexographic Gravure

Printing processes on different materials

Process	Advantages	Disadvantages	Applications
Lithography	Good reproduction quality especially photographs Cheap printing process Able to print on a wide range of papers High printing speeds Widely available	Colour variation due to water/ink mixture Paper can stretch due to dampening	Business cards, stationery, menus, brochures, posters, magazines, newspapers
Letterpress	Dense ink gives good printing quality Less wastage of paper than other processes	High cost Slow process	Books with large amounts of text, letterheads and business cards
Gravure	Consistent colour High speed Ink dries on evaporation Good results on cheaper paper	High cost of printing plates and cylinders Only good for long print runs	High quality art and photographic books, postage stamps, packaging, expensive magazines
Screen printing	Economical for short runs Stencils easy to produce Can print on virtually any material	Difficult to achieve fine detail Low output Print requires long drying time	T-shirts, posters, plastic and metal signage, point-of-sale displays
Flexography	High speed Relatively cheap to set up Can print on same presses as letterpress	Difficult to reproduce fine detail Colour may not be consistent	Less expensive magazines, paperbacks, newspapers, packaging

Printing processes

Enhancing the format of paper and board

Aims

- To understand a range of techniques used for enhancing the format of paper and board.
- To understand the processes used for providing visual impact to products.

Die cutting

Die cutting is a machine process which involves punching out a specified area of the sheet of paper or board for dramatic effect. The die is a sharp metal blade which can be manufactured to any shape in order to cut the appropriate shape or hole. Die cutting is widely used to round the corners of paper as seen on playing cards.

When die cutting is used in conjunction with folding processes, the results can be very dramatic indeed.

Folding

Producing a small brochure involves trimming and folding. The most common folding methods for a brochure are:

- single-fold (four-page brochure)
- accordion (eight-page brochure)
- gate-fold (eight-page brochure)
- roll-over styles (eight-page brochure).

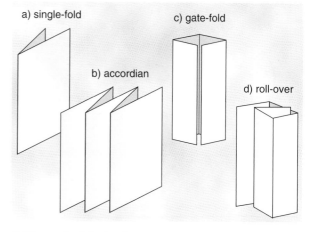

a) single-fold
b) accordian
c) gate-fold
d) roll-over

Folding methods for brochures

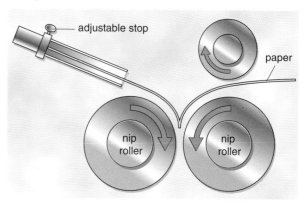

adjustable stop
paper
nip roller
nip roller

The process of buckle folding

The most widely used machine for folding paper is a buckle folder. The paper enters the machine and is stopped at the required distance using an adjustable stop. The paper is then scored at the fold line by two inward-revolving rollers called nip rollers.

Prehistoric Animals - Dinosaurs weren't the only animals to roam the earth in prehistoric times. Prehistoric animals include all living things that inhabited the earth from the beginning of life about 3½ billion years ago until about 5,000 years ago when humans first began to keep written records.

Saber-toothed Tiger - Saber-toothed tigers first appeared some time between 35 and 36 million years ago during the period called the Oligocene epoch. The scientific name for the saber-toothed tiger is smilodon. This group included many different types of cats, all of which had long canine or sharp upper jaws. These cats wore razor sabers (canines), and used their sharp teeth to pierce the thick skins of plant-eating animals, such as mammoths.

One of the best known saber-toothed tigers is the Smilodon, which means "knife tooth" in Latin. It lived in South America about 2 million years ago during the Pleistocene epoch. Smilodon was twice as long and its large grew to over 1 inches in length, it probably looked a bit like a modern-day female lion.

Die cutting and folding can be used to produce dramatic effects

■ Hints and tips ■

When producing a brochure, the paper may often crack or fold poorly unless it is scored first. **Scoring** with a scoring tool or pair of scissors will produce a clean, crisp fold and improve the presentation and quality of your brochure.

Laminating

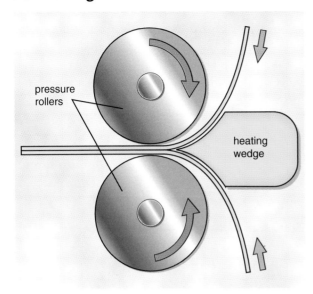

The lamination process

Lamination adds strength and durability and gives a high gloss finish to paper or card. Film lamination is glued to the paper as it is fed through a heating wedge under high pressure.

The lamination of paper or card gives:

- added protection
- an enhanced surface finish (either gloss or matt finishes are available).

Polypropylene (PP) is the most widely used plastic for laminating, although PET can be used for applications where the product has to be more durable and scuff resistant.

Varnishing

Varnishing is another method of providing a high-gloss coating to paper or board. The two main processes used are ultraviolet (UV) and aqueous (water-based).

Ultraviolet varnishing

This process uses UV radiation to harden a liquid plastic coating applied to the surface of paper or board. It is the ultimate in high-gloss coating that can be applied during the printing process.

Spot UV varnishing can be applied to 'jazz up' certain areas of a printed piece that has already been matt-film laminated. This produces highlighted areas for visual impact.

Aqueous varnishing

This process uses a water-based varnish providing a gloss coating that can either be applied during or after the printing process. It is not, however, as high gloss as UV varnishing.

Embossing

Embossing is a process by which areas of paper or board are raised above the surrounding area. It is used to provide a sophisticated look to the covers of many books or packages.

An embossing die can be made to any required shape or size and consists of a raised image or letters. The die is simply pressed into the paper or board under pressure to create the embossed effect.

Hot foil blocking

Hot foil blocking transfers a foil coating to paper by means of a heated die. A roll of foil with a polyester backing sheet is continuously fed over the paper or other material and a heated die presses the foil on to the surface of the paper.

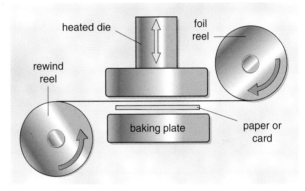

The hot foil blocking process

Hot foil blocking gives a quality effect to paper and card and is often used in packaging to add visual impact.

Hot foil blocking for visual impact

∎ Things to do ∎

1 Collect examples of each enhancement technique described in this section.

2 Discuss the reasons why a manufacturer may use expensive enhancement techniques on their products.

Preparing, processing and finishing materials

Preparation and manufacture

Aims

- To understand that standard components are available to designers.
- To understand how commercial nets can easily be adapted to suit individual needs.

In essence, it is sometimes not the job of the designers to 're-invent the wheel' and they should take into account standard components already available.

Designers are often given the brief to come up with **innovative** designs that don't look like anything else available on the market at that time. However, the majority of design solutions will incorporate existing designs, technologies or components in order to reduce development costs.

Standard components are available commercially for designers to browse through catalogues and pick out the best component relating to a particular design brief. These components are tried and tested solutions to needs and are immediately available for manufacture.

The distinctive slim 250ml aluminium can seems to be the standard component for energy drinks

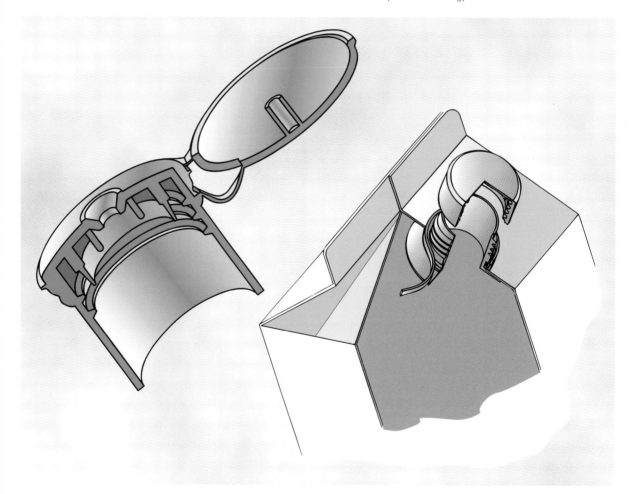

Standard closure components for packaging such as the integral hinge flip-top cap (left) and the resealable screw-top closure (right) are available commercially, so the designer does not have to 're-invent the wheel'

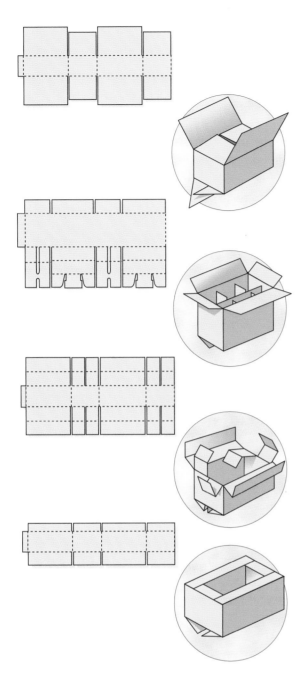

Commercial packaging nets

Many companies produce standard packaging nets for designers to adopt and simply to add individual **graphic identity**. There is also an international system using diagrams of net constructions and symbols to avoid the need for long and complicated verbal descriptions. This is of great benefit as regardless of the language spoken, the references may be used in orders and specifications for packing cases worldwide.

Adapting commercial packaging nets

Standard packaging nets can be used as a starting point. It is quite simple to adapt the dimensions and proportions of these to suit individual needs. For example, a student developing a **point-of-sale** dispenser for a new sweet container researched a commercial packaging net for such a purpose (see diagram).

The student then made several scale **prototype** models of the net in order to work out the most suitable size for the new product (see photo). In this way, the student was able to save time by not having to draw out the required net from scratch.

Standard commercial packaging nets shown as diagrams avoid the need for lengthy descriptions

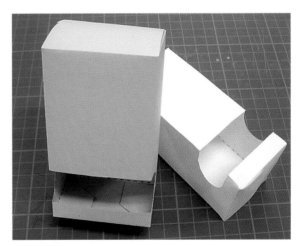

Simple paper models exploring size and proportion

■ Things to do ■

1 Using an existing packaging net as a start, adapt its dimensions to create a different sized package.

2 Write to companies asking for information on the standard components they manufacture, i.e. bottles closures and packaging nets.

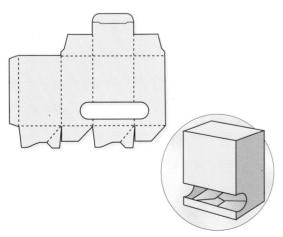

Commercial packaging net – autolock bottom with gravity dispenser

Preparing, processing and finishing materials

41

Testing prototypes

Aims

- To understand anthropometric and ergonomic considerations when designing 3D products.
- To understand the importance of testing mock-ups and models to aid designing.

Anthropometrics

Anthropometrics is the name given to the study of human physical dimensions in relation to the objects which are used by people.

The aim of anthropometrics is to improve:

- safety
- convenience
- quality of life.

In order to achieve these, studies have been made into the various dimensions of the human body for different age groups including height, width, length of reach and even size of fingers. The data obtained from these studies has been published by the British Standards Institute (BSI). It is worth including such data in your design projects.

Ergonomics

Ergonomics is concerned with the relationship between people and the products they use. Ergonomics makes use of anthropometric data to ensure that products can be used comfortably by the people for whom they are designed.

Designing for people

When designing a product for people to use, such as a computer mouse, mobile phone or hand-held radio, it is useful to produce a number of **mock-ups**.

For example, a student wishes to design and produce a prototype model of a hand-held radio preset to a radio station that could be given away as a promotional item. Here it is extremely important that the designer gives consideration to ergonomic design as the product is held in the hand.

A student's foam model of hand-held radios to test ergonomic considerations

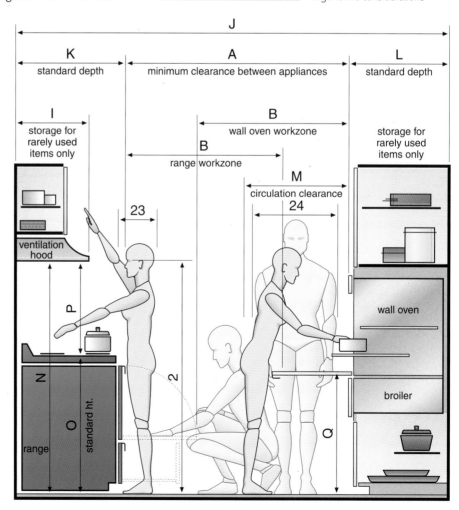

RANGE CENTRE

	in	cm
A	48 min.	121.9 min.
B	40	101.6
C	15	38.1 min.
D	21–30	53.3–76.2
E	1–3	2.5–7.6
F	15 min.	38.1 min.
G	19.5–46	49.5–116.8
H	12 min.	30.5 min.
I	17.5 max.	44.5 max.
J	96–101.5	243.8–257.8
K	24–27.5	61.0–69.0
L	24–26	61.0–66.0
M	30	76.2
N	60 min.	152.4 min.
O	35–36.25	88.9–92.1
P	24 min.	61.0 min.
Q	35 max.	88.9 max.

An example of anthropometric data

The size and contours of the hand must be explored through a series of mock-up models in order to determine a comfortable hand grip. The designer can test the effectiveness of the design through 3D modelling rather than relying on a 2D drawing. A foam model can easily be reshaped in order to best-fit the hand.

The professional product designer

Before any commercial product is put into production it is modelled at every stage of its development. When Microsoft designed a new computer mouse especially for Internet use, the product had no rubber ball underneath like a typical mouse. A scroll wheel was used instead to allow the user to scroll up and down and an optical sensor fitted to the underside allowed tracking from side to side. The job of the design team was to design a distinctive looking mouse that reflected this new technology.

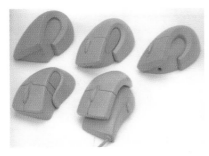

Initial designs were based around the shape of a seashell, not just for aesthetics but based upon

Prototype models exploring comfortable shapes for the new mouse

the results of extensive ergonomic studies. A number of foam models were produced to test how comfortable the designs were to use.

As a result of testing these early models, the design team found that they had to modify some aspects of the original design. The large swooping curves of the original were smoothed slightly. A less dramatic, but still distinctive shape emerged. Further models were produced to determine colour schemes and surface finishes – extremely important if the new mouse was to appeal to its target market.

The final Microsoft IntelliMouse Explorer has a distinctive design that not only looks good but is comfortable to use. The optical sensor's red LED light is clearly visible glowing underneath the mouse which increases its appeal.

The Microsoft IntelliMouse Explorer

Paper and card modelling

Paper or card modelling is a quick and simple way to test the effectiveness of design ideas. For example, when designing a pop-up card or promotion, it is highly recommended that a number of models be produced. In this way, the pop-up mechanism can be explored and tested for its visual impact. Don't forget to record your findings!

A student's card models for pop-up mechanisms

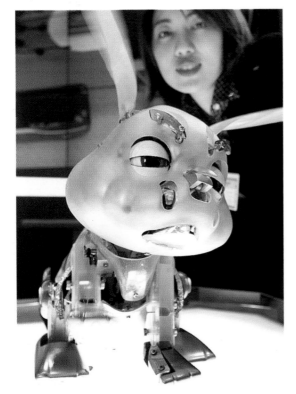

At the annual Tokyo toy show, Japanese toy giant Bandai unveils a prototype model of a robot pet called 'Patata', which can react to human voices and is able to make facial expressions

■ **Things to do** ■

1 Look up relevant anthropometric data for your projects at your local library.
2 Use simple foam mock-ups or paper models in your design and make activities in order to test the effectiveness of your designs.

Cutting processes

Aims

- To be able to select the correct tools and equipment for cutting different materials.
- To understand laser cutting as an industrial process for cutting and engraving plastics.

Over the course of your GCSE studies in Graphic Products, you will design and make 2D and 3D products. Therefore, you may move from a graphics studio working mainly in paper and card into a workshop area where you will use wood and plastics.

It is extremely important that you take care when using any tool or piece of equipment. Make sure that you take the necessary precautions and identify any potential hazards.

Laser cutting plastics

Sign-makers will often make use of the accuracy and fine finish of lasers in cutting out and engraving sheet materials. Lasers are computer controlled so that signs can be designed directly on screen with no loss of quality or definition when outputted to the laser cutter/engraver. As a result, lasers are capable of working to more precise **tolerances**.

Tool/equipment	Task	Tool/equipment	Task
Scissors	Cutting paper, thin card and thin plastic sheet, i.e. acetate, thin vacuum forming plastic (HIP)	Rotary trimmer	Cutting/trimming accurate straight lines in paper, thin card and plastic sheet
Scalpel	Cutting paper, card and thin plastic sheet and foam board	Scoring tool	Scoring lines in paper and card prior to folding
Craft knife, i.e. Stanley knife	More heavy-duty knife for cutting paper, board, HIP and foam board	Safety rule	Used in conjunction with scalpel/craft knife to cut materials
Circle cutter	Cutting circles in paper, thin card and thin plastic sheet	Cutting mat	Provides a self sealing surface to protect desk-top when cutting with scalpel/craft knife

Cutting tools/equipment for the graphics studio

Tool/equipment	Task	Tool/equipment	Task
Twist drill	Cutting tool for drilling holes in most materials	Hot wire cutter	Heated wire for cutting expanded and rigid foam
Pillar drill	Bench mounted electric drill for drilling holes in most materials. Used in conjunction with twist drills, etc.	Surform	Shaping soft materials such as laminated MDF models
Hand drill	For drilling holes with twist drills by hand	File	Creating a smooth finish on acrylics and MDF
Tenon saw	Sawing straight lines in sheet materials such as MDF and pine	Needle file	Smoothing edges in very fine and precise work
Coping saw	Cutting curves out of sheet materials in both thin wood and acrylics	Glasspaper	Making a smooth finish on woods
Abrafile	Cutting curves in thin sheet metals and acrylics	Wet and dry paper	Making a smooth finish on plastics
Vibro saw	Bench-mounted electric saw for cutting thin sheet materials in wood and plastics	Emery cloth	Making a smooth finish on metals
Bandsaw	Heavy-duty electric saw for cutting most sheet materials. For teacher and technician use only		

Cutting tools/equipment for the workshop

The most versatile material for laser cutting and engraving is acrylic. One of the reasons for this is that acrylic keeps its highly polished shine when engraved. Acrylics also offer an excellent edge quality when cut with a laser, resulting in a glass-like polished edge when using clear acrylic.

■ **Things to do** ■

1 List all the cutting equipment you would use to produce:
   a a pop-up greeting card
   b a styrofoam mock-up model
   c an MDF model of a computer games console
   d a plywood architectural model

2 Use the Internet to find information on the laser cutting of acrylics.

Joining processes

Aims

- To select the correct adhesive for different jobs.
- To recognize the hazards of solvent-based adhesives.

Joining processes can be categorized as follows:

- Permanent – once made they cannot be reversed without causing damage to the work piece (i.e. welding metals)
- Temporary – although not always designed to be taken apart, they can be disassembled if needed without causing damage (i.e. knock-down furniture joints)
- Adhesives – substances for bonding two or more materials together using a chemical reaction (i.e. glueing acrylic using Tensol cement)

Joining materials using adhesives

In most graphic product projects, a 2D or 3D model will be made. This means that there will be two main categories of adhesives: those for joining compliant materials such as paper and boards and those for joining resistant materials such as wood and metal.

The correct selection of adhesives for different purposes is extremely important and should be considered carefully.

Health and safety

Many adhesives are solvent based containing volatile organic compounds (VOCs) which give off vapours that can make you feel ill if inhaled. Therefore, these adhesives are extremely hazardous to use and should only be applied in well-ventilated areas (with extrac-

Adhesive	Uses	Properties
Polyvinyl acetate (PVA)	Woods	Sold in a plastic container as a white, ready-mixed liquid. Easy to apply and sets in two hours. Gives a strong joint. It is not waterproof
Epoxy resin	Expanded polystyrene to most materials	Sold in two tubes – resin and hardener which are mixed together in equal quantities. Chemical reaction hardens immediately, but full strength achieved after two to three days
Contact adhesive	Metals and plastics; dissimilar materials, i.e. plastic to wood and metal to wood General purpose; fabric to most materials	Usually used for glueing sheet materials. Sold in a metal tube for easy application
Acrylic cement	Acrylic to acrylic	Solvent based for rapid bonding of acrylics. Sold in a screw-top metal can and needs a separate applicator. Always read manufacturer's instructions
Polystyrene cement	High impact polystyrene to high impact polystyrene	Solvent-based adhesive that melts surface of pieces to be joined and causes them to weld together. Able to use a brush to apply adhesive to surfaces and absorbed into joint using capillary action
Hot melt	Most materials for rapid joining (usually temporary)	Used with a glue gun. Solid glue stick passes through a heating element and becomes a gel. Easy to apply from gun nozzle. Does not give a solid joint – more of a temporary fixing
Glue stick	Paper, card and board	Solvent-free and non-toxic adhesive available in an easy-to-use solid stick. Can cause paper to crinkle if not applied evenly
Artwork spray	Paper, card and board	Repositionable adhesive sold in aerosol form. Excellent for presentations as work does not wrinkle and allows for repositioning
Double-sided tape	Most materials, especially paper, card and board	Tape with adhesive on both sides. Quick and easy to cut to size and peel off protective cover. Excellent for mounting work for exhibitions

Adhesives for sheet and modelling materials

EVO-STIK ADHESIVE CLEANER

For removing surplus adhesive. Before using, ensure Cleaner does not attack substrates. Test a hidden area first.

EVO-STIK ADHESIVE CLEANER

Contains Toluene. Harmful by inhalation. Keep away from sources of ignition including pilot lights – No Smoking. Keep container tightly closed in a well ventilated place. Keep out of reach of children. Take precautionary measures against static discharges. Do not breathe vapour. Use only in well ventilated areas

HIGHLY FLAMMABLE

HARMFUL

Evode Ltd., Common Road, Stafford, England.

PLEASE TAKE THIS CARD TO THE CHECKOUT WITH YOUR OTHER PURCHASES. THE CASHIER WILL SUPPLY YOU WITH THIS PRODUCT.

A Joint Campaign by EVO-STIK and this store to help prevent solvent abuse.
Thank you for your co-operation.

Display card for purchasing a solvent-based adhesive

tion, if possible, but always with windows and doors open) and a respirator or mask should be worn. Under the Intoxicating Substances Supply Act 1985, it is an offence to supply, or offer to supply a solvent to a person under 18 years old. Many retailers now only display a card on their shelves which you have to take to the check-out in order to purchase the adhesive.

Some adhesives can be corrosive which means they can burn if they come into contact with your skin. It is important that you wash the area thoroughly with warm soapy water immediately. If the adhesive splashes in your eye, then consult your designated first aid person who will use an eye bath to wash the area.

In all cases, always read the manufacturer's instructions before using any adhesive.

COSHH

The Control of Substances Hazardous to Health (COSHH) regulations state that the storage of dangerous materials such as solvent-based adhesives be in a secure and easily identified metal cupboard. Manufacturers are also required by law to provide a safety data sheet with their products that outline any potential risks to users.

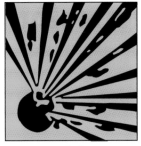
explosive

highly flammable

corrosive

toxic

harmful

oxidizing

Manufacturers' warning symbols

Developments in adhesives

In 1999, scientists in the USA developed some new forms of adhesives which are based on sugar, rather than oil-based epoxies. Currently, in standard glues epoxy groups form polymers with petrochemicals. In the new development, epoxy groups are attached to sucrose molecules instead of a petrochemical. By varying the constituents, many new adhesives can be created. These have the potential to be environmentally friendly since sugar is a renewable source.

■ **Things to do** ■

1 Experiment with various adhesives to join a range of scrap pieces of material.

2 Discuss the potential risks of solvent abuse.

Thermoforming plastics

Aim

- To understand the main industrial processes for thermoforming plastics.

Many items are mass produced using thermoforming processes to mould and shape plastics. Thermoplastics are commonly used because they can be moulded very easily and any waste produced can be recycled and used again in the process.

Injection moulding and **blow moulding** can produce large quantities of identical products very quickly. Vacuum forming is also widely used in industry and is ideal for producing batches of similar products within schools.

Injection moulding

Injection moulding is the most widely used industrial method of producing moulded plastic products. The process involves plastic granules being fed from a hopper into a heated barrel by an Archimedean screw

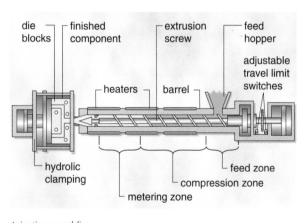

Injection moulding

Snap-on covers for mobile phones injection moulded from ABS plastic

and forced under pressure into a mould. Once the product has been formed, the mould opens and the product is ejected. Modern injection moulding is fully automated and involves very little labour, especially as the product produced has a high surface finish and needs little finishing by hand.

Blow moulding

Blow moulding is widely used to produce hollow objects such as bottles of all shapes and sizes. A small hollow tube called a parison is extruded (stretched) downwards between two halves of a mould. The mould then closes around the neck and the base and hot compressed air is blown into the parison. The parison expands out on to the walls of the mould and forms the bottle shape. Once cool, the newly formed bottle is ejected from the mould.

Blow moulding can produce very intricate mouldings and is especially useful for producing accurate screw threads for the necks of bottles.

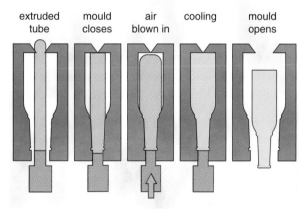

Extrusion blow moulding

■ Hints and tips ■

You can tell if a product has been blow moulded by a feint line running down each side and across the bottom where the two halves of the mould have come together.

Vacuum forming

Vacuum forming is probably the easiest method of **thermoforming** plastics in most schools. To vacuum form you first need to make a mould, usually out of MDF because it is easy to shape and has a good surface finish.

The **mould** is placed on a platen on the vacuum former and lowered to sit above a vacuum cavity. Thermoplastic sheet (usually high impact polystyrene – HIP) is securely clamped above the mould and slowly heated under the heater.

Once the plastic is soft, the platen is raised and a switch starts the vacuum forming process. A vacuum is created underneath the mould and atmospheric pressure pushing down on top of the mould causes the plastic to pull over the mould.

The same process is used in industry but on a much larger scale with multiple moulds.

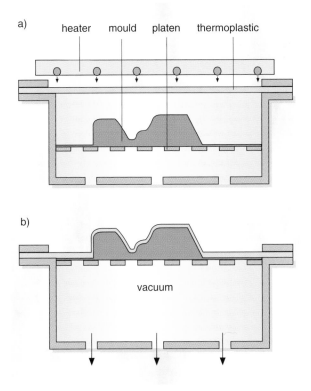

Vacuum forming

Just one example of the many dairy-based products using vacuum forming in their packaging

■ Things to do ■

1 Study a selection of plastic products and try to determine what forming process was used in their manufacture.

2 Design and make a mould for vacuum forming a batch of identical products.

Process	Advantages	Disadvantages	Plastics used	Applications
Injection moulding	Ideal for mass or continual production – low unit cost for each moulding for high volumes. Precision moulding – high quality surface finish or texture can be added to the mould	High initial set-up costs as mould expensive to develop and produce	Nylon, ABS, PS, HDPE, PP	Casings for electronic products, containers for storage and packaging
Blow moulding	Intricate shapes can be formed. Can produce hollow shapes with thin walls to reduce weight and material costs. Ideal for mass or continual production – low unit cost for each moulding	High initial set-up costs as mould expensive to develop and produce. Large amounts of waste material produced	HDPE, LDPE, PET, PP, PS, PVC	Plastic bottles and containers of all sizes and shapes, i.e. fizzy drinks bottles and shampoo bottles
Vacuum forming	Ideal for batch production – inexpensive. Easy to make moulds. Moulds can be easily modified	Mould needs to be accurate to prevent 'webbing' from occurring	Acrylic, HIP, PVC	Chocolate box trays, yoghurt pots, blister packs

Plastic forming processes

Lay planning

Aim

- To understand the need for efficient use of materials using lay planning.

■ **Hints and tips** ■

When marking out on any material, make sure that you mark up to a square edge.

The basic principle

When you mark out any work on a sheet of material, it is important that you mark out correctly in order to:

- save material
- save costs
- save time.

Look at the examples of marking out below.

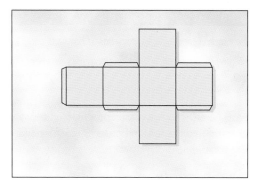

Inefficient marking out results in large amounts of waste

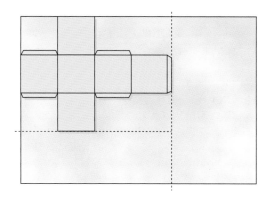

More efficient use of material means less waste

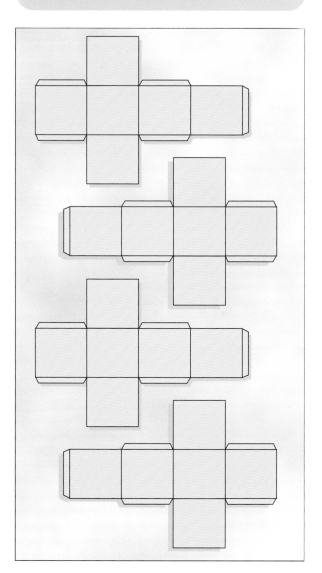

Marking out multiple nets on a sheet of material

Efficient lay planning

Lay planning, widely used in the textiles industry for marking out and cutting patterns, can be applied to the packaging industry and your projects as well.

Where multiple copies of a packaging net are needed from a sheet of material, efficient lay planning is essential to keep wastage to a minimum, thereby saving money. It is therefore important to calculate how many nets can fit on to a single sheet of material allowing for each net to be cut out accurately.

Nowadays, most of these calculations are carried out by computers using sophisticated mathematical programs. The **net** can be drawn on screen and the computer will automatically work out the most efficient method of laying a number of nets on a specified piece of material.

Reducing material use in industry

The packaging industry is full of examples where reducing the amount of materials used to manufacture a product is both cost effective and environmentally aware.

Aluminium drinks cans

During the process of manufacturing the top of a drinks can, the circular shapes need to be stamped out of a coiled sheet of aluminium. Careful calculations are made to limit the amount of aluminium used.

a)

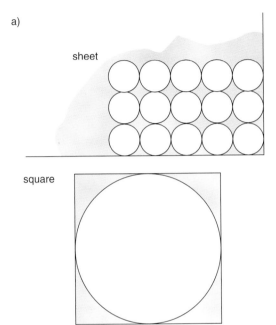

Here the tops are in a square formation, each sitting in its own square of aluminium

b)

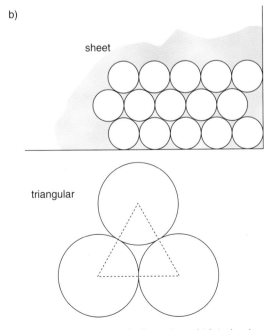

Here the tops are in a triangular formation, which is the closest that they can be packed together

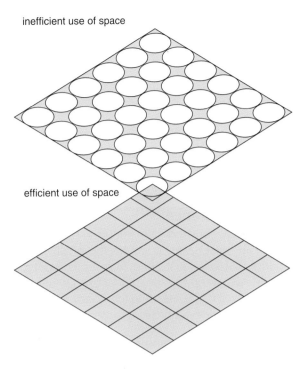

Using space

When the scrap aluminium for each is worked out, the triangular formation produces just 9.3 per cent scrap compared to a staggering 21.4 per cent scrap for the square formation. It is obvious from this example that by using efficient lay planning less aluminium is wasted and is therefore more cost effective.

Efficient use of space

The efficient use of space is an important aspect of any commercial package. The shape of the pack can determine how many of the product can be displayed on the supermarket shelf and how many can be transported at one time. For instance, a square box can be stacked easily and therefore more can be displayed in a small area and more boxes can be loaded into a lorry for delivery. This ensures that the product is always on display in the supermarket and that fuel costs are cut due to fewer deliveries.

> ■ **Things to do** ■
>
> 1 Using squared grid paper, work out the most efficient use of material for a number of basic nets for a box.
>
> 2 Make up a number of different shaped nets and see how they can be efficiently packed for storage and display.

Quality of manufacture

Aims

- To understand the need for quality assurance procedures when manufacturing products in quantity.
- To understand the importance of working to specified tolerances.

Quality assurance

Why do people buy a certain product in preference to others? In most cases, it will be the quality of the product that will set it apart from its competitors. Usually, we want to buy products that are well designed, well made, are value for money and will last us a long time without breaking.

Quality assurance is concerned with monitoring the quality of a product from its design and development stage, through its manufacture to its end-use performance and degree of customer satisfaction. All companies will use quality assurance and the most successful ones are awarded the ISO 9000, an international standard of quality.

Total quality management

Total quality management (TQM) is the goal of companies which seek to establish the highest possible standards of quality at all stages:

- Design – the development of a high quality product that will appeal to its target market.
- Manufacture – the 'build quality' of a product using quality materials, manufacturing processes and assembly.
- Performance – the product's fitness-for-purpose or how well it works.
- Customer satisfaction – value for money, reliability and after-sales service.

In addition to these stages, TQM is also about a company's attitude to its workers. Each department within a company is treated like a customer with the same high standards of service and attention shown when dealing between departments.

Quality control

Quality control is part of the achievement of quality assurance. It is concerned with monitoring and achieving high standards by inspection and testing.

Inspection

Inspection is the examination of the product or component in order to check that it is within a specified **tolerance**.

Tolerance is the degree to which a component is acceptable in order to function in accordance with its design specification. Tolerance is expressed as an upper (+) and lower (−) limit deviation from the ideal size (+ / −).

For example, the diameter of a bottle must be such that it fits in the bottle-filling machinery at the bottling plant and also contains the required amount of liquid. A 54 mm diameter bottle is likely to have a dimensional tolerance of + / − 0.8 mm. If, when inspected and tested, a bottle measures between 53.2 mm and 54.8 mm, it would be within the agreed tolerance and would therefore be accepted. Any bottle that lies outside this tolerance will be scrapped.

Obviously, it is far too time consuming to inspect every single component that is made, so inspection is carried out on selected samples (i.e. one in every 100). The results are then recorded on quality control charts that can be used to identify whether a machine is producing components inside of the tolerance limits. This is extremely important because if a machine were to be left unchecked it could produce components that were outside of tolerance and would be wasted. As a result of quality control inspection, the machine can be adjusted with minimum disruption to production saving both time and costs in wastage.

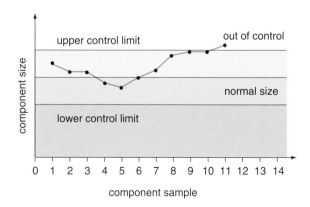

Quality control chart

Testing

Testing is concerned with the product's performance. Tests are carried out in laboratory conditions with strict control procedures to ensure that the results obtained are accurate. Tests are carried out on the final product, individual components and the materials used. Testing can be divided into two categories:

- Non-destructive testing – where the product is tested until there are signs of cracking, etc. to determine how much force is needed to deform it. Alternatively, components can be examined using X-rays to spot the tiniest of defects.
- Testing to destruction – where the product is destroyed to see how it reacts, for example car collision tests where data can be obtained to help develop safety.

Testing materials aims to find out how they react to forces. For example, paper is tested to destruction to work out its tensile strength (how much it can be pulled until it tears) by clamping one end and stretching the other. In the photo below, a paper tissue is being tested on a machine to see what happens when it is dispensed. Obviously, you do not want your tissue to tear when it is pulled out of its box.

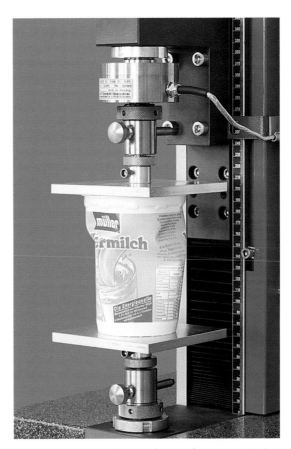

Testing the plastic used in a yoghurt pot by compressing it

Other tests try to find out a material's strength under compressive forces (a squashing force). A material that will easily buckle when stacked for transportation and display will be of little use.

British standards

The British Standards Institute (**BSI**) is responsible for setting standards, testing products and monitoring quality assurance for companies in the UK and overseas. This independent organization is responsible for testing a wide range of products and agreeing a set of standards by which all future products are tested and evaluated. Any product that meets a British Standard is awarded a Kitemark, as long as the manufacturer has quality systems in place to ensure that every product is made to the same standard.

> ### ▪ Things to do ▪
>
> 1 Conduct a set of tests on different types of paper to find out their tensile strength.
> 2 Conduct a set of tests on a yoghurt pot, aluminium drinks can and a small cardboard box to find out their compressive strengths.

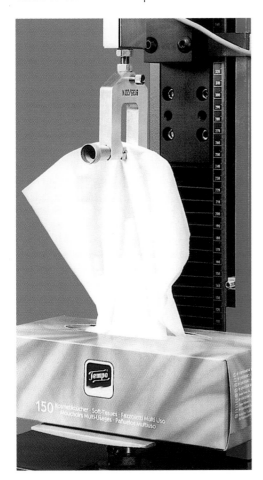

Testing paper tissue for tensile strength

Health and safety

Aims

- To understand the legislation concerning health and safety in all workplaces.
- To identify hazards in making activities and to carry out the appropriate risk assessment.

The Health and Safety at Work Act 1974 states that measures should be taken wherever possible to safeguard the risk of injury to employees (or students). A school workshop, for example, must have fire extinguishers, a clearly marked fire exit and everyone should be aware of the correct procedures upon discovering a fire and evacuating the building.

All machines should carry safety signs to alert users to any potential risks while operating. You should also be familiar with the safety signs asking you to wear safety goggles on all machinery.

Stop machine before removing guards

Fire exit

Wear glasses

Fire extinguisher

Standard safety signs found in a workshop

Government guidelines for health and safety issues within any workplace, including schools, are laid out by the Health & Safety Executive (HSE). The HSE states that all places of work must carry out a risk assessment of their facilities. This is in order to identify any potential hazards to employees (or pupils) and to put in place any control measures to reduce the risk of injury.

The HSE suggests that there are 'Five Steps to Risk Assessment':

- Identify the hazard.
- Identify the people at risk.
- Evaluate the risks.
- Decide on control measures.
- Record assessment.

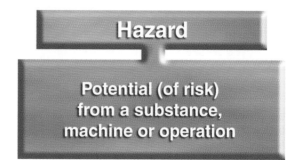

Hazard

Potential (of risk) from a substance, machine or operation

Risk

Reality (of harm from the hazard)

Control measure

Action taken to minimize the risks to people

What is the difference between a hazard and a risk? (Source: Adapted from the HSE diagram)

Using a scalpel for cutting

There are certain points you must remember when using a scalpel or sharp knife of any kind. To avoid injury:

- Always cut away from your body.
- Always pay full attention to what you are cutting.
- Use a safety rule to protect your fingers.
- Use a cutting mat to protect the cutting surface.

Below is a typical risk assessment for using a scalpel.

Hazard	People at risk	Risk	Control measures
Using a scalpel for cutting	Individual (yourself)	Cutting fingers or hand open with an extremely sharp scalpel blade	Always cut away from body Always pay full attention to what you are cutting Use a safety rule to protect fingers Use a cutting mat to protect surface

Risk assessment for using a scalpel

Health and safety does not only apply to making activities, but designing activities as well. For example, when using computers you will sit in front of a VDU (visual display unit) screen for quite a long time. The

HSE together with the European Union 'VDU Directive' has regulations and guidance on working at a computer workstation.

■ Hints and tips ■

If using a computer at home for long periods, make sure that you take regular breaks in order to rest your eyes from the flicker of the VDU screen.

■ Things to do ■

1 Identify any hazards in your design and make activities and compile a risk assessment table.

2 Enter this information into a spreadsheet and use it in your project work.

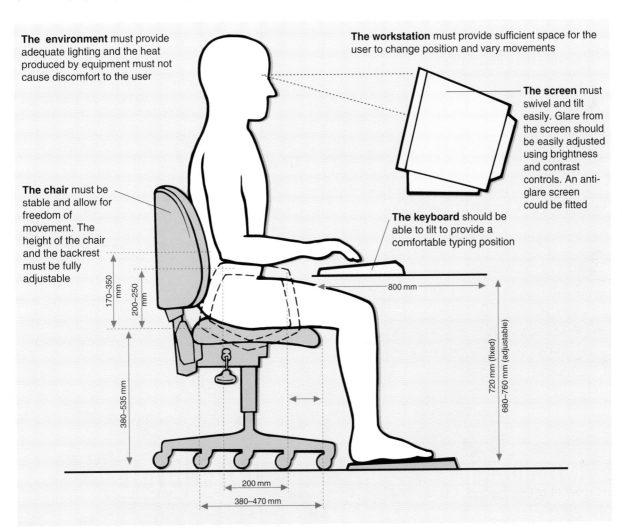

The environment must provide adequate lighting and the heat produced by equipment must not cause discomfort to the user

The workstation must provide sufficient space for the user to change position and vary movements

The screen must swivel and tilt easily. Glare from the screen should be easily adjusted using brightness and contrast controls. An anti-glare screen could be fitted

The chair must be stable and allow for freedom of movement. The height of the chair and the backrest must be fully adjustable

The keyboard should be able to tilt to provide a comfortable typing position

170–350 mm
200–250 mm
380–535 mm
800 mm
720 mm (fixed)
680–760 mm (adjustable)
200 mm
380–470 mm

Working safely and comfortably at a computer

Computer-aided design

Aims

- To understand the use of computers as a design tool for modifying designs.
- To understand the use of computer-aided animation in the media.

Computer-aided design (CAD) can be used for producing illustrations or for 3D modelling. All you need is a good computer with a drawing package and image manipulation software if using scanned images. A colour printer will enable computer-generated work to be quickly printed out for proofing.

Digital illustration

The use of computers for producing illustrations for books, magazines, packaging, flyers and other promotional material is now widespread.

The professional illustrator

The work of Mark Oliver is a typical example of how a professional illustrator using mainly traditional media such as paints has moved towards digital illustration using computers.

Mark produces high quality cartoon-like illustrations for many companies and was commissioned by the confectionery manufacturer Trebor Bassetts to redesign the packaging of its sweets.

How Mark creates digital illustrations

1 An initial drawing of the illustration is given to the client for approval.

2 Once approved, the rough is scanned and then imported into a sophisticated illustration software package.

3 The next step is to trace over the scanned image using computer software. This creates shapes and lines that the computer can recognize and is a very skilled process.

4 Using the illustration package, the line drawing is coloured using flat colours. The end result is a digital illustration!

Other professional designers remove the need for a rough sketch using pen or pencil and construct their illustrations directly on screen.

Advantages of using CAD for modifying designs

- Designs can be modified on screen.
- Alternative colours and lettering styles can be explored.
- Designs can be stored and easily retrieved.
- Design data can be directly outputted to printers or computer-aided manufacture (CAM) machinery.
- Multiple copies can be made from design data.

3D modelling

Using 3D modelling can give a more realistic impression of a product than a flat 2D image. The use of sophisticated software enables the designer first to construct a wire frame model on screen. This model can be viewed in all directions by the use of different 'camera' angles.

Mark Oliver's digital illustrations for 'Bassetts and Beyond'

The wire frame model can then be rendered or given the appropriate surface finish or texture. This involves wrapping a 'skin' around the wire frame to give a photo-realistic image of the product if it were to be put into production.

Other uses of 3D modelling can be seen clearly in the development of computer games. Here 3D wire frame models are constructed and rendered for backgrounds or animated to provide the game play. Animated graphics can also be seen on TV as title sequences for shows or commercials and as special effects in films.

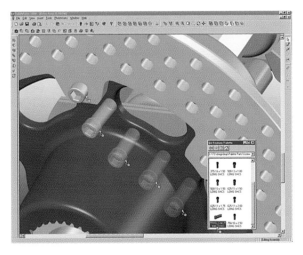

3D CAD model

Channel E4 commercial

London-based Aldis Animation was commissioned by Channel 4 to create a TV commercial for the E4 entertainment channel.

First, it created a simple storyboard around the idea of E4 logos escaping from a sardine can. All of the individual elements for the commercial were then created digitally using wire frame models. These models were animated on screen and rendered using sophisticated software packages. Finally, lighting effects and different camera angles were added to create the finished commercial.

The entire project took just two weeks to complete because of the use of computers as opposed to traditional animation processes.

■ Things to do ■

1 Make a list of all the computer-generated sequences you see when watching TV one evening. Is the style or format the same for different programmes?

2 Describe sequences from recent films that have incorporated the use of digital special effects.

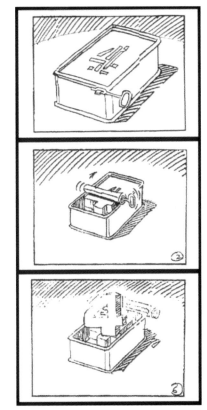

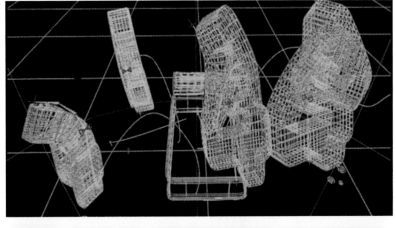

Creating a TV commercial using 3D modelling – from storyboard to wire frame models to rendered animation

Designing with desktop publishing

Aims

- To understand the basic process of designing using desktop publishing software.
- To understand that desktop publishing can be used to produce high quality documents.

■ Hints and tips ■

DTP packages are ideal for producing graphic products that look professional. These include brochures, business cards, headed paper, CD covers, flyers, nets for packaging, etc.

Desktop publishing (DTP) combines the features of word-processing, graphic design and printing in one package. Modern newspapers and magazines all use DTP to produce the layout of their pages electronically rather than the traditional process of typesetting. The two main DTP packages used in most large companies are QuarkXPress and Adobe PageMaker.

There are four main stages in the production of a document using DTP.

Stage 1

The basic text is either word-processed using a standard word-processing package (for example Microsoft Word) and can be either cut and pasted or flowed into the DTP package, or typed directly into the DTP package.

Stage 2

Illustrations or graphics are created on the computer using a graphics package or by tools within the DTP package itself. Photographs or other images can be scanned into the computer, manipulated on screen using packages such as Adobe Photoshop, and imported into the DTP package.

Stage 3

The word-processed text and graphics are combined within the DTP package. A page layout grid can be created to produce columns for lining up text and graphics. At this stage, text and graphics can be manipulated or cropped to best fit the style and layout of the document.

Designers working on papers and magazines will create multiple grid layouts to organize quite complicated pages.

Stage 4

A **proof copy** of the finished document is printed using an output device such as a colour printer or laser printer. This proof copy can be evaluated and any changes can be made on screen.

Multiple grid layouts

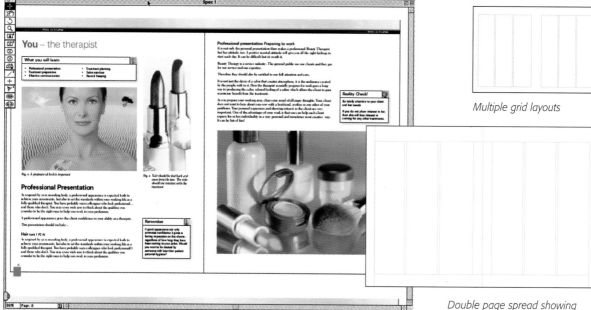

Pages created in QuarkXPress

Double page spread showing a three column grid

Advantages of desktop publishing for modifying designs

- Text and image manipulation
- Wide availability of typefaces
- Graphic tools available
- Page layout grid and guides can be easily added on screen (invisible when printed)
- Clip art library for illustrations and diagrams
- Cut and paste facility to alter page layouts
- Ability to zoom in and out for attention to detail

- A house style template can be created for producing documents in a standardized format (for example headed paper)
- Standard templates available for different sized brochures, business cards, CD-ROM labels, etc.

■ Things to do ■

1 Using a page from a magazine, draw in the grid the designer may have used to construct the page.

2 Using a template on a DTP package, produce a business card promoting yourself as a graphic designer. Local companies may need business stationery producing and are willing to pay for new designs.

An excellent example of how DTP can be used to incorporate text, graphics and scanned images

Input devices and resources

Aims

- To understand the range of input devices and resources available to the designer in order to produce high quality documents.

Input devices

Input devices are used to capture images and convert them into a format that can be recognized by the computer.

These are the common formats for saving images:

- **PICT File** – an abbreviation for picture format. One of the earliest file formats for saving pictures.
- **TIFF – Tagged Image File Format** is the standard format for saving graphic images. It has a very clear resolution and is therefore ideal for DTP work.
- **JPEG – Joint Photographic Experts Group** is a compressed file ideal for use for images published on the Internet. A certain amount of resolution is lost, however, when the file is decompressed, for example the image may pixelate when pasted into a document and re-sized or stretched.

Scanners

These allow the designer to input already existing images such as a picture from a magazine, into a computer.

A **scanner** uses three coloured beams (red, green and blue) to scan the image line by line measuring the colour and brightness at points along each line. By repeating the process point by point and line by line, the scanner builds up a complete representation of the image. The smaller the distance the scanner scans between each point, the higher the resolution and the clearer the picture. High resolution images, however, take up a lot of memory space on the computer.

Digital cameras

These allow the designer to take photographs and directly input them into a computer without the need for film processing. A **digital camera** converts light into electrical signals and stores them as pixels (dots).

Low and high resolution scanned images

A flatbed scanner

The more pixels the camera stores, the higher the resolution of the picture, and the clearer the image. Once the image has been captured it is stored on the camera's memory card or a floppy disk. The image can then be downloaded on to a computer for printing or saved into a document.

Resources

There are a number of resources available in order to produce a professional looking document.

Clip art libraries

Most DTP packages include a clip art library that can be used to cut and paste images into a document. The use of an appropriate clip art image or photo can enhance any page. There is, however, a tendency to use too many or the wrong clip art images. Clip art is available in a wide range of topics and styles that are copyright free.

CD-ROMs

There is a wide range of **CD-ROMs** available to help you when designing. Some, such as encyclopaedias, are largely information based and you can use these to help you with your research. Others are design packages for specific purposes such as interior or garden design. Design-based CD-

Students using a digital camera at school

ROMs contain a wide range of specific symbols, templates and design tools in order to help you create exciting designs.

Note: ROM stands for Read Only Memory which means that information on the disk can only be read by the computer and you cannot change any of the information on the disk.

The Internet

The Internet is a rich source of images that can be cut and pasted into a document. Certain websites contain interactive pages that can help to explain a difficult concept or idea, or provide important information. By typing in a key word related to a project, a number of relevant websites will be listed for you to investigate.

Examples of clip art images

■ **Hints and tips** ■

Saving your work
- Always save your work in the correct place on your school's system.
- Saving an image from the Internet: always cut and paste it into a DTP or similar package. Do not save it as a stand-alone PICT file as it will take up a lot of memory.

■ **Things to do** ■

1 Use a digital camera to record the stages in the disassembly of a product or the manufacture of a product.

2 Produce a leaflet on a subject of your choice involving the cutting and pasting of images from clip art and the Internet into your document.

3 Scan an image into the computer and manipulate the image on screen using the appropriate software.

Computer-aided manufacture

Aims

- To understand the use of computer-aided manufacture in single-item production.
- To understand the use of computer-aided manufacture for rapid prototyping in industry.

Computer-aided manufacture (CAM) can be a useful tool in producing high quality graphic products in schools. There are a number of relatively simple software packages available which will help you to design a product on screen and then output to a machine for making.

For example, a student has set the design brief: *'To design and prototype a toiletries gift set'*. Part of the student's design ideas focused on the development of a bar of soap and a promotional key ring, both involving CAM in their manufacture.

A logo to bring a brand identity to the range of toiletries was designed using a relatively simple 2D software package on a computer which could be connected to either a plotter-cutter or an engraving machine.

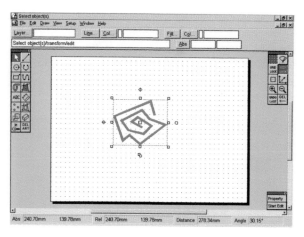

A logo was designed using 2D software

First, the logo data was sent to the plotter-cutter in order to produce a vinyl sticker. The sticker was placed on a shaped piece of acrylic to form a simple but effective key ring.

A vinyl sticker, produced on a plotter-cutter, was placed on a shaped piece of acrylic to make this keyring

Next, a template was made on screen to fit the exact sizes of another piece of clear acrylic to be used in the manufacture of the bar of soap. The logo was re-sized to fit the template and a slogan added for effect. The data this time was sent to an engraving machine which engraved the logo and slogan to a depth of 3 mm.

Engraving using CAM

Several more layers of clear acrylic were laminated together and smooth wet 'n' dry paper was used to produce the effect of a bar of soap. The end result was a simple prototype of a bar of soap with an embossed effect.

Using CAM, this simple prototype of a bar of soap with an embossed logo and slogan was produced.

3D models. The process is based on the computer slicing the 3D object into hundreds of very thin layers (0.125–0.75 mm thick) and transferring the data from each layer to the laser.

The laser draws the first layer of the shape on to the surface of the resin which causes it to solidify. The layer is supported on a platform which moves down to enable the next layer to be drawn. This process of drawing, solidifying and moving down quickly builds up one layer on top of the other until the final 3D object is achieved.

Most companies do not have this technology. Instead, they use rapid prototyping services from specialist companies. Stereolithography prototypes can typically be delivered within three to five days of the customer's design data being received, therefore saving both time and money.

Rapid prototyping

The need for manufacturing industries to cut down on the time and costs involved in developing a new product has led to the development of rapid prototyping. This involves the creation of 3D objects using laser technology to solidify liquid plastic polymers (resins) in a process called **stereolithography**.

Using sophisticated software that a stereolithography machine can read, 2D CAD drawings are converted to

■ Things to do ■

1 Use CAM in your design activities where appropriate.

2 Identify components in your design activities that could be manufactured using CAM.

3 Look on the Internet and in a business directory for your area to locate rapid prototyping services.

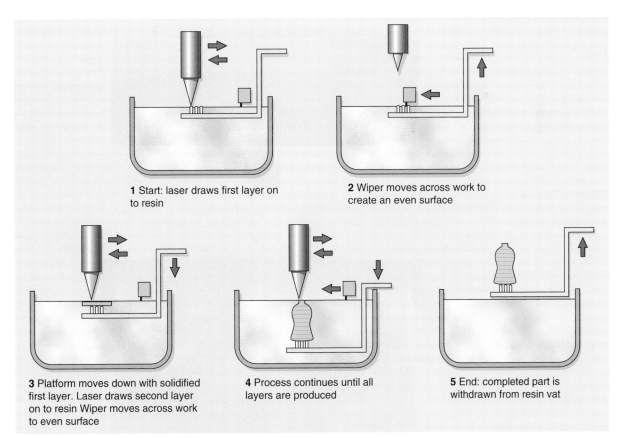

1 Start: laser draws first layer on to resin

2 Wiper moves across work to create an even surface

3 Platform moves down with solidified first layer. Laser draws second layer on to resin Wiper moves across work to even surface

4 Process continues until all layers are produced

5 End: completed part is withdrawn from resin vat

The process of stereolithography

Output devices

Aims

* To understand a range of output devices suitable for copying computer-generated designs.

As we have seen, designing using computers allows us to store data that can then be transferred to an output device such as a colour printer. This means that data can be retrieved at any time and multiple copies of the same design can be made. There is a range of output devices available in many schools.

Printers

A printer is an output device that prints text or graphics on to paper, thin card and film, such as overhead projector transparencies and coated glossy papers.

Inkjet printers

An inkjet printer is capable of producing low-cost, high quality text and graphics in colour by combining the colours blue (cyan), red (magenta), yellow and black (CMYK). Inkjet printers use thermal technology to produce heated bubbles of ink that burst spraying ink at the surface of a piece of paper to form the image. The printhead prints in strips across the page, moving down the page to build up the complete image with a resolution that can range from 300 to 1200 dpi (dots per inch). They are sometimes referred to as 'bubble jet' printers.

Advantages

* Inexpensive to buy and relative low cost of printing per page.
* High quality, full-colour images at high resolution.

Disadvantages

* Ink cartridges have to be changed frequently.
* Expensive coated papers are needed for photo quality images.
* Colours can 'bleed' on low quality papers.
* Ink takes a little time to dry, therefore images can smudge.
* Printer nozzles may become blocked.

EPSON
STYLUS
PHOTO 750

An inkjet printer can produce high quality colour graphics

Laser printers

These printers are available in black and white or colour and use the same technology as photocopier machines. The printer receives data from the computer, which is processed and used to control the operation of a laser beam directing light at a large roller or drum. By altering the electrical charge, wherever the laser beam hits the drum creates the required image. The drum then rotates through a powder called toner. The electrically charged areas attract the powder and the print is made when it is transferred on to the paper by a combination of heat and pressure.

Advantages

- Very high quality text and graphics are produced.
- High printing speeds (up to 20 text pages per minute).

Disadvantages

- Expensive to buy and replace toner cartridges (especially colour).
- Most film and coated papers cannot be used due to heat generated.

Plotters

A **plotter** is a high quality output device that draws an image on paper or any other suitable medium directed by commands from a computer.

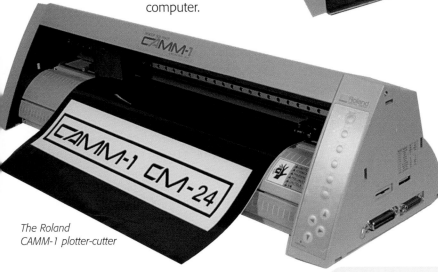

The Roland CAMM-1 plotter-cutter

XY plotters

The term XY refers to the axes along which the plotting pen can travel. Plotters differ from printers in that they can draw lines using a pen which results in continuous lines. Printers can only simulate lines by printing a series of closely spaced dots that appear to form a line. Therefore, plotters are used for engineering drawings where precision is needed.

Plotter-cutters

Plotter-cutters can plot drawings in the same way as XY plotters, but they can also produce cut shapes in card, vinyl and other thin sheet materials using thin blades. The blade can be adjusted for depth and pressure of cut by commands within the computer software. This allows the plotter-cutter to undertake finely controlled cutting techniques such as 'scoring' in which the card that is to be folded is only cut to a certain depth to make folding easier. Plotter-cutters are used in the production of advertising and promotional products as well as in sign making and similar processes.

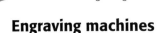

The Roland CAMM-2 engraving machine

Engraving machines

Engravers also have a variety of uses, ranging from sign making to the production of jewellery and 3D reliefs. Industrial-sized engravers can operate in the X, Y and Z axes. This allows the engraving of 3D surfaces and curves as well as lettering. Cheaper engravers commonly available in schools have less control over the Z axis of the cutting tool meaning that it can only cut at one depth.

■ Things to do ■

1 Make full use of your school's output devices to produce batches of graphic products.

2 Compare costings for producing multiple copies of your designs in your projects including printing from printers at school or home and printing from commercial printers.

Practice
examination questions

1 a The table below shows some items of equipment. Complete the table by:

    i naming each item of equipment **(3 marks)**

    ii giving a suitable description of the use for each item of equipment. **(3 marks)**

The first is done for you.

Equipment	Name	Use
	Hot melt gun	Tool for glueing rigid materials

b Give three safety procedures that must be followed when using the hot melt glue gun. **(3 marks)**

c Name the most suitable adhesive for the following uses:

Use	Suitable adhesive
Glueing paper on to paper	
Glueing a photograph on to card (allowing for repositioning)	
Glueing corrugated card on to foam board	

(3 marks)

d The tabs on a cereal packet are either tacked (spot glued) or securely fixed using a suitable adhesive.

    i On the net below, clearly label the tabs that will be tacked and those that will be securely fixed. **(3 marks)**

    ii Give reasons for your choice. **(3 marks)**

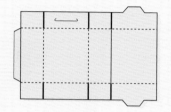

2 A colour leaflet promoting the Valley Leisure Centre is to be distributed in the local area.

a The leaflet is to have a gloss surface finish to the paper.

    i Name two printing effects that could be used to give a gloss finish. **(2 marks)**

    ii Explain why a gloss finish to the paper may be needed. **(2 marks)**

b 20 000 copies of the leaflet are to be printed. Name a suitable printing process and describe two advantages of using this process.

(4 marks)

3 a Give two reasons why computers are popular/useful for word processing. **(2 marks)**

b Name two different uses of computers when designing a graphic product. **(2 marks)**

c Describe two qualities computers have that makes them suitable for small-scale production. **(4 marks)**

Section C:
Manufacturing commercial products

One-off production

Aims

- To understand one-off production processes.
- To understand the advantages and disadvantages of one-off production.

One-off production is where a single item is required, often in response to an individual customer's need or requirement.

Historical background

Before the Industrial Revolution, products were often designed and made by the same craftsperson. For example, the village blacksmith was highly skilled and produced the majority of metal-based products for the local community. His skills and knowledge were passed down from generation to generation and this is how it remained for centuries.

Modern one-off production

The modern craftsperson is still very skilled, but will often work on jobs designed by other people. For example, a company needing a vacuum-formed tray for a cosmetics gift box would require a mock-up of the final product before it could consider the box for full production. A pattern maker may be commissioned to make a wooden mould of the tray and vacuum form it to test its fitness-for-purpose.

The role of the pattern maker is increasingly becoming taken over by the use of **CAD/CAM** and rapid prototyping systems. Using relatively simple computer software a 3D computer model of a mould can be constructed. The data for this mould can then be downloaded into a computer controlled machine such as a router which would cut the shape of the mould. Although the quality and accuracy of the final mould would be similar to that of the craftsman, the time taken to produce it would be drastically reduced.

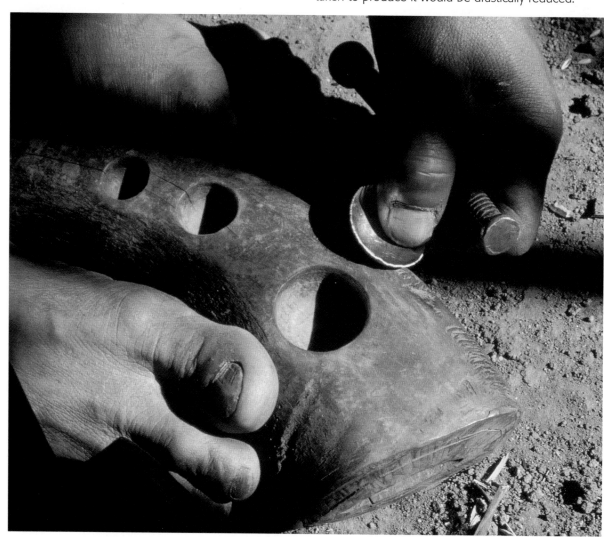

In a northern Thailand village, a silversmith uses a mould made from buffalo horn to make the pieces for a silver head-dress

The noted Swiss architect, urbanist and interior designer known as Le Corbusier stands behind an architectural model

With this in mind, manufacturers will often use computers to aid in the development of a product rather than the expensive craftsperson.

Model makers

Specialized model makers are commissioned to make architectural models or interior layouts for proposed buildings. A highly detailed 3D model can communicate much more about the building than any 2D drawing ever could.

Model making is a highly skilled job that requires great attention to detail and is not easily achieved by the use of computers. Because of this, both design and production costs are much higher.

Model making degree courses are available at a number of Universities. During these courses, students are trained to make models for a range of commercial applications i.e. starships or aliens for science fiction movies, shop displays for retailers or medical teaching aids.

Advantages of one-off production

- Product made to exact specifications of client
- Often paid for on completion of commission.

Disadvantages

As only one product is being made, the cost of one-off production tends to be much higher because of the:
- quality of materials used
- intensive and highly skilled labour required
- design and production costs.

As a result, all design and development costs have to be recovered from the sale of a single product.

■ Things to do ■

1 Make a list and sketch examples of products using one-off production.

2 Make a simple card model of your school buildings to a scale of 50:1.

Custom-made vinyl graphics

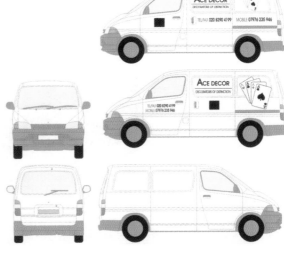

Aim

- To understand the processes involved in producing one-off vinyl graphics for a client need.

Sign-makers' shops are commonly found on the outskirts of large towns or on industrial estates. When a company needs graphics for promotion it will often approach the design team in such a shop.

Visuals for a client's work van

Design

The client's needs will be discussed with the designer and a number of 'visuals' created on computer.

For vehicle graphics, the sign-maker has a large database of current vehicles on computer to match those used by the client. The turnaround for visuals will be between one and two days and the client will then decide which one he or she wants to use.

In addition to the actual design, the client also has a choice of colours and quality of vinyl. For example, high quality vinyl for external use such as vehicles will last for about ten years, whereas internal shop displays may be made from a thinner vinyl designed to last for around three years.

HIGH PERFORMANCE CAST VINYL FILM *Series 2100/2200 7 Year Durability*

00 CLEAR	63 VIOLET	04 GOLD METALLIC	89 AQUA BLUE
67 PRIMROSE YELLOW	75 LAVENDER	84 EGGSHELL	71 TEAL
06 YELLOW	74 ROYAL PURPLE	26 OYSTER	53 DARK BAHAMA BLUE
97 LIGHT ORANGE	99 BRIGHT PURPLE	68 BEIGE	98 SLATE GREEN
83 BRIGHT ORANGE	62 PURPLE	48 CAMEL BEIGE	70 PALM OYSTER GREY
96 TANGERINE	92 WEDGEWOOD BLUE	27 FAWN	81 TITANIUM
73 POPPY	51 ARUBA BLUE	94 SANDUST CLAY	28 DOVE GREY
14 TOMATO RED	31 OCEAN BLUE	78 CORAL	80 GRAPHITE
60 BRIGHT CARDINAL RED	18 OLYMPIC BLUE	85 COCOA	50 MEDIUM GREY
42 CARDINAL RED	32 AZURE BLUE	08 BROWN	07 SILVER METALLIC*
01 RED	37 MEDIUM BLUE METALLIC	35 BROWN METALLIC	21 SMOKE METALLIC
12 BURGUNDY	05 BLUE	79 KIWI	52 DARK GREY
39 BURGUNDY METALLIC	17 SAPPHIRE BLUE	69 APPLE GREEN	23 CHARCOAL METALLIC
91 SALMON	66 MEDIUM BLUE	61 KELLY GREEN	40 MATT BLACK
95 BRIDAL ROSE	30 DARK BLUE METALLIC	106 MEDIUM GREEN	03 BLACK
76 MAGENTA	93 LIGHT NAVY	24 DARK GREEN	49 MATT WHITE
58 DARK MAGENTA	65 MIDNIGHT BLUE	89 FOREST GREEN	02 WHITE
56 HOT PINK	11 DARK BLUE	87 AQUA GREEN	
82 RASPBERRY	59 IMITATION GOLD	88 MINT GREEN	

*Indicates reduced outdoor durability: See Product Information Bulletin for details.

A sign-maker's colour swatch

Manufacturing commercial products

Manufacture

As in the production of vinyl stickers in a school, the digital image is sent to the vinyl cutter for cutting. However, in our sign-maker's example, the software used is more sophisticated and the vinyl cutter is much larger and able to cut larger widths of material.

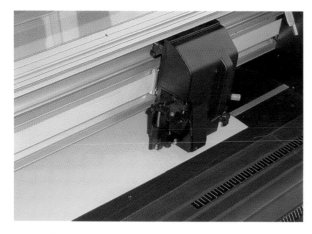

Cutting the vinyl

If the client is a large, well-known company, it is likely to have a digital version of the company logo available since it is important that the logo looks the same wherever it appears. Clients can e-mail this data to the sign-maker directly so eliminating the need for a disk.

Once the cutting is complete, all of the background vinyl is taken away in a process called '**weeding**'. This leaves only the required graphic.

'Weeding' away the background vinyl

Finally, a roll of application tape is applied over the graphic. The application tape will stick to the vinyl graphic so that it can be removed from the backing sheet and on to the sign or vehicle. Once applied the application tape can be removed leaving only the vinyl graphic.

'Squeegeeing' down the application tape

The end results can be very dramatic giving real impact to any promotion. Vinyl graphics can be found on all types of signage from shop fronts to the side of works vans or even on racing cars.

Racing car vinyl graphics

▪ Things to do ▪

1 Why are vinyl graphics used instead of the traditional hand-painted method?

2 Note down and sketch or photograph examples of vinyl graphics in your local area and use them as resource material for your design activities.

Batch production

Aims

- To understand the batch production process
- To understand the characteristics of a flexible manufacturing system within batch production.

This level of production makes products in specified quantities. They can be made in one production run of thousands or just ten items depending upon the scale of the project. **Batch production** lies between needing more than a one-off product but less than the large-scale manufacture of mass-production processes.

Flexible manufacturing systems

Flexible manufacturing systems (**FMS**) are intended for batch production work where the time taken to set up new tools and machines for different jobs is crucial. The aim is to make the most efficient use of manufacturing processes while remaining flexible enough to respond quickly to the needs of other jobs.

In the late 1960s all the leading Japanese manufacturers moved towards the flexible factory to stay competitive. With the increase in global competition and the creation of open markets such as in Europe, consumers are demanding an increase in the range of products they can buy. The flexible factory, involving batch production processes, will use more flexible equipment that has the ability to perform more than one task on a wide range of products.

Characteristics of flexible manufacturing

A flexible manufacturing system will involve the use of computers both to plan and manufacture products including:

- **CAD/CAM**
- project management software
- computer-integrated manufacture (**CIM**)
- computer numerically controlled (**CNC**) machining centres.

All of the above will be explained in the next section of this book.

Books are batch produced

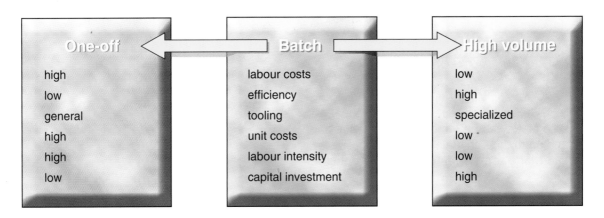

One-off	Batch	High volume
high	labour costs	low
low	efficiency	high
general	tooling	specialized
high	unit costs	low
high	labour intensity	low
low	capital investment	high

Batch within the scales of production

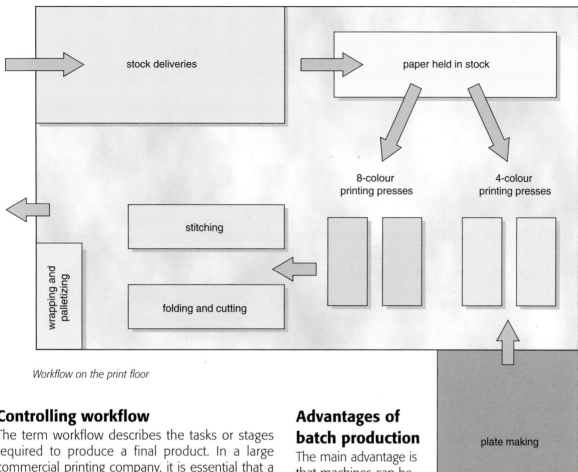

Workflow on the print floor

Controlling workflow

The term workflow describes the tasks or stages required to produce a final product. In a large commercial printing company, it is essential that a range of printed materials can be produced for different customers at the same time.

In the example below, the printing company uses two 8-colour lithographic printing presses and two 4-colour presses. The various presses are being used for different print jobs at any one time. For instance, the 8-colour presses may be printing a batch of 5000 full-colour brochures for one customer while the 4-colour presses are involved with the mass production of a leaflet for another customer.

The print floor is organized in such a way that paper stock is transformed into the final printed product as efficiently as possible in the shortest amount of time. Large stock deliveries supplied on lorries enter the factory via a loading bay and using a forked-lift truck, pallets of paper stock are unloaded and stored.

Printing plates are designed and made and put on to the printing presses prior to printing. The paper is taken out of stock and fed into the printing press for full-colour printing. Once printed, the paper is transferred to the finishing area where it is folded, stitched and trimmed. When ready, the final printed products are counted and put on pallets ready for delivery or transfer to a warehouse for storage.

Advantages of batch production

The main advantage is that machines can be reset once a batch has been produced, then used to produce a batch of another product. This allows:

- rapid response to customer demands
- the batch size to be increased or decreased while in production.

Workers who operate this type of machinery are likely to be more skilled because of the versatility of the machines and tools.

Disadvantages of batch production

The main disadvantage of batch production comes when appropriate planning is not carried out and serious problems occur, for example high storage costs while the product is being sold.

■ Things to do ■

1 Set up a production line for the batch production of a simple product, for example a simple decorated box or greetings card.

2 Why would a large sportswear manufacturer such as Nike use large scale batch production for the manufacturer of a new sports shoe?

Batch production in the printing industry

Aims

- To understand that full-colour images are made up of four process colours
- To understand four-colour process printing using colour separation.

Designing for a client

Many commercial printing companies will have their own design department which will work with the client on a design brief. Alternatively, the client may have his or her designs already planned and a digital version is given to the printer. Either way, discussion between the designer and client will determine the budget, scale and quality of the final printed materials produced.

How overlapping dots create a full colour image

Four-colour process printing

Four-colour process printing is the method used by commercial printing companies to reproduce a full-colour image. The four colours used to print an image are cyan (blue), magenta (red), yellow and black (**CMYK**). These four are called the process colours. In order to print a full-colour piece of work the colour image must first be 'separated' or broken down into its four process colours. A separate printing plate for each colour can then be produced. This is known as colour separation.

Colour separation

Colour separation is achieved either using a laser **scanner** or, using the latest technology, a printing plate can be made directly from a computer. In the case of an electronic laser scanner, the original full-colour image is taped to the drum of the scanner which then rotates at high speeds.

The scanning drum in action

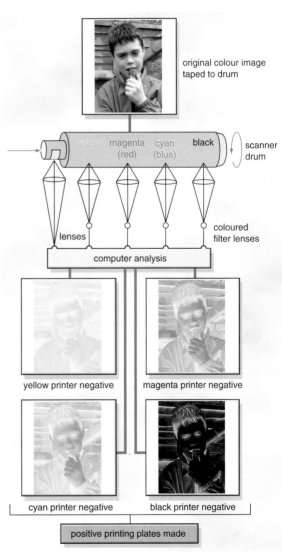

The colour separation process

The scanner uses special coloured filters in order to separate the process colours from each other. A screen is also used to convert the separated colours into dots. These dots of cyan, magenta, yellow and black when overlaid, one on top of the other, are what make up the full-colour image.

Once the process colours have been separated, the scanner is then able to produce four printer negatives from which four different printing plates can be made.

Colour proofs

Before the job is printed, the printing company produces a sample of the printed piece for checking. This can either be a proof made from the printing plate itself or, at an earlier stage, a pre-press proof showing the four colour separations on transparencies. The inks used are transparent so that when the four separated colours are printed, they blend to give the full-colour effect.

Using the latest technology, a colour proof can be produced directly from a computer using an IRIS printer. This printer will reproduce the final colours extremely accurately and is often used for presentation to the client. Any minor adjustments can be made on computer at this stage.

Making a printing plate

For lithographic printing, a thin aluminium printing plate with a photo-sensitive coating is used. The process is very similar to that of developing and etching a printed circuit board using ultraviolet (UV) rays.

The finished plates are inspected for quality. If under a magnifying glass there are any scratches or dirt found, then the plate is scrapped and another made.

For accuracy and speed, some printing companies are able to send the colour separation information directly from a computer to a hi-tech platemaker where the printing plates are etched using a series of lasers. This also reduces the human handling time on the plates and so reduces the possibility of damage.

Colour printing

There are several ways of printing in full colour. These include **offset litho**, **letterpress**, **gravure** and **silkscreen** (see pages 34–6). In the case of large printing companies, 8-colour lithographic printing is often used. Here the paper can be printed on one side with four colours, turned over and four colours printed on the other side. This is achieved in one run and is capable of producing around 10 000 double-sided pages per hour.

Finishing

Most publications will contain several printed pages. In the case of saddle-stitching, the pages are folded on a folding machine and fed along a conveyor where another folded page is added. The collated pages are then wire stitched along the spine using wire staples and passed on to a series of guillotines for trimming. The completed publication is put on pallets and shrink wrapped prior to delivery to the customer.

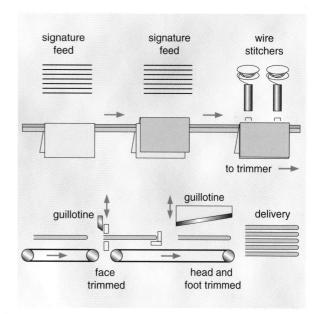

Folding, binding and trimming

■ Things to do ■

1 Visit a local printing company to witness the process first hand.
2 Look closely at bill board posters to see the overlapping dots used to create the full colour image.

Continuous production

Aim

- To understand the processes used in the manufacture of a drinks can.

The manufacture and filling of a drinks can

More than 8.5 billion drinks cans are manufactured in the UK each year, and over 30 billion in total throughout Europe. The manufacture and filling of a drinks can is carried out 24 hours a day and is therefore an example of continuous production.

stage 1 making the can

stage 2 making the can end

stage 3 the filling process

Stage 1: Making the can

1 Huge coils of aluminium or steel arrive at the can manufacturing plant. The metal is coated with a thin film of oil to act as a lubricant.

2 The metal is fed continuously through a cupping process which blanks and draws thousands of shallow cups every minute.

3 Each cup is drawn through a series of ironing rings to form the walls of the can. Trimmers remove the surplus irregular edge.

4 The trimmed can bodies are washed and dried to remove all traces of oil.

5 The clean cans are coated with a base coat of lacquer which forms a good surface for the printing inks.

6 The cans pass through a hot air oven to dry the lacquer.

7 A highly sophisticated printer/decorator applies the printed design in up to six colours, plus a varnish, using a dry-offset litho printing process.

8 The cans pass through a second oven which dries the inks and varnish.

9 The inside of each can is sprayed with lacquer to protect the can from corrosion.

10 Once again, lacquered internal and external surfaces are dried in an oven.

11 The cans are passed through a necker/flanger so that the can end can be attached.

12 Every can is tested at each stage of manufacture. At the final stage it passes through a light tester which automatically rejects any cans with pinholes or fractures.

13 The finished can bodies are transferred to the warehouse where they are automatically put on pallets before despatch to the filling plant.

Stage 2: Making the can end

1 A lubricated coil of aluminium is fed through a press which stamps out thousands of ends every minute. At the same time the edges are curled.

2 The newly formed ends are passed through a lining machine which applies a very precise bead of compound sealant around the inside of the curl.

3 A video inspection system checks the ends to ensure they are all perfect.

4 The pull tabs are made from a narrow width coil of aluminium which is pierced and cut to form the tab.

5 The ends pass through a series of dies which score them and attach the tabs, which are fed from a separate source.

6 The final product is the retained ring-pull end.

7 The finished ends are packaged in paper sleeves and put on pallets for shipment to the can filler.

Stage 3: The filling process

1 The can bodies and ends are removed from the pallets. The machine operator registers the codes from pallets to ensure that they can be traced back to source if needs be, at a later date.

2 Cans are then mass combined and sent to the filling machine at speed using high tech air conveyors.

3 The cans are turned upside down and are cleaned using high pressure air and water.

4 Once clean, the cans are turned the right side up again.

5 The cans are then immediately passed into a covered filler section, ensuring no further chance of contamination.

6 All air is extracted from the can by passing them through a gassing system where they are filled with CO_2 (carbon dioxide – the gas that puts the 'fizz' into drinks).

7 Up to 2000 cans per minute are then filled with the drink.

8 The cans leave the filler and pass directly into a seamer where the can ends are fed from a separate source.

9 All remaining air is displaced through a further injection of CO_2 and immediately mechanically sealed.

10 An interlocking seam is formed at high speed. Up to 2000 cans are processed each minute.

11 Cans then pass through a detector where any incorrectly filled cans are rejected.

12 The filled cans go through a coding process where details of the filling date and 'best before' date are printed on the base of the can.

13 The cans are then put into their multi-pack format.

14 Finally, the cans are shrink-wrapped, put on pallets and despatched to the distributor or retailer.

Adapted from *The Can Makers.*

■ Things to do ■

1 Identify the stages in the manufacture of a can where quality control is used. Why is quality control necessary at these points?

2 What do you suppose happens to all of the waste metal produced by some of the manufacturing stages?

3 Find out about the production of another product using continuous production.

High-volume (mass) production

Aims

- To understand the development of mass-production processes.
- To understand how production lines are used to provide large quantities of identical products.

Large quantities of products are produced, sometimes using continuous production, 24 hours a day.

Historical background

Mass production dates back to the Industrial Revolution in the mid-1700s and the invention of the steam engine for powering machinery. The rapid development of machinery and transport meant that there was a need for new products that could no longer be crafted by individuals. The craftsperson worked by hand, but products were usually time consuming to make and expensive to manufacture and buy.

The traditional craftsperson became specialized and was employed by large factories to produce only a part of the overall product. This became known as **division of labour** and, as a result, production processes became simpler. As tasks became ever simpler, machines were introduced to improve efficiency (**mechanization**) and there was little need for the craftsperson any longer.

At the beginning of the twentieth century, the American Henry Ford, who recognized the importance of the motor car in modern life, set about improving production processes at his factory to increase efficiency and improve output. Ford introduced assembly lines and subdivided tasks so that manufactured components or parts were delivered to the workers to be assembled, rather than the workers having to spend time moving about the factory to obtain the parts they needed.

Mass production and the use of assembly lines made production easier and faster, enabling products to be made more cheaply than ever. Highly specialized machinery was used and was therefore expensive, but the vast amounts of products produced ensured that the cost of each product was kept low. This meant that less well-off people could now afford to buy many products that previously only the rich could afford.

Modern mass production

Industrial technologies and industrial thinking have changed considerably since World War II (1939–45), particularly over the past 25 years. This has been due to the rapid development of computers and communications technology. Within the developed world, the manufacturing industry has become dependent upon computers.

Systems and control

Modern mass-production processes involve the use of many systems in order to manufacture products, including:

- in-line production and assembly
- production cells
- just-in-time (JIT) stock control.

In-line production and assembly

This is considered to be the traditional approach to mass production. It utilizes low-cost, unskilled labour supported by semi-skilled, flexible people able to change tasks as required to ensure smooth running of the assembly line.

Production cells

Production cells operate as separate units within a large manufacturing plant. Each cell produces a specific component which is then fed into the larger manufacturing system which could be an assembly line or another cell. The individual cell has responsibility for every aspect of the production of the component including quality control and maintenance.

Ford's Model-T assembly line

A production cell

It is extremely important that members of the cell work together as a team and share responsibility for the cell output. Production cells vary from small teams of people to fully automated cells with computer-controlled machines and robots.

Just-in-time

Traditionally, products have been manufactured according to an agreed schedule or plan that is governed by the availability of materials and components. Just-in-time (**JIT**) was developed in Japan and is based upon production being 'pulled' along in response to customers' orders and the requirements of the manufacturing process.

The aim is to rid the system of stock at every opportunity. Stock includes materials and components and finished products held in storage. This wastes investment (money) and takes up valuable space in the factory. JIT enables manufacturing systems to be flexible (respond quickly to changes), but it is totally dependent upon the supply of materials and components arriving just in time. Suppliers of bought-in components such as bottle caps are often bound by contracts that result in fines if they are not able to supply the assembly line as required.

Productivity

Modern mass-production systems ensure a company's high productivity – that is, the speed and efficiency with which a company turns raw materials into a final product. The most common measure of productivity is the amount of products a single worker can produce. The more products that a single worker can produce, the lower the labour costs. If labour costs are kept to a minimum, then the higher the potential profit will be.

Advantages of mass production

The main advantage of mass production processes is that unit costs are lower due to increased efficiency. This may also result in increased sales. Other advantages include:

- specialization – work is divided into specific tasks with labour to match the job
- increased production means that set-up costs are quickly recovered
- bulk buying of raw materials at lower prices.

Disadvantages of mass production

- High costs due to buying and setting up of large-scale machinery.
- Social implications, that is, an unskilled and bored workforce who simply 'mind' the machinery or complete repetitive tasks all day.

Type of company	Weekly wage	Output per worker production	Labour cost per unit of
Efficient	£280	40.0	£7
Less efficient	£250	25.0	£10

Comparison of productivity

■ Things to do ■

1 Find out about the mass-production processes involved in producing a range of graphic products.

2 Imagine that you are a worker on an assembly line – what would be the advantages and disadvantages of your job?

Computer-integrated manufacture

Aims

- To understand the system of computer-integrated manufacture.
- To understand the sub-systems within a computer-integrated manufacturing system.

Computer-integrated manufacture (**CIM**) is an inter-linked network of computers controlling machinery and the flow of information during the manufacturing process. The system is totally automated and involves the control of sub-systems such as **CAD/CAM** and **CNC** machines.

All aspects of a company's operations are integrated so that different computers can share the same information and communicate with one another. Computers are used to link production systems and business information with manufacturing operations in order to create cooperative and smooth running production lines. The tasks performed within CIM will include:

- the design of components
- planning the most effective production stages and workflow
- controlling the operations of machines
- performing business functions, i.e. ordering stock and materials and invoicing customers.

Controlling workflow

Project management software allows a manufacturer to create different workflows for different types of jobs and organize production cells in an overall production schedule.

For example, within a single company, a **CAD** file might be sent automatically from the designer to a production engineer for comment, then on to the purchasing department for materials. At each stage in this workflow, one individual or group is responsible for a specific task, for example making sure that the right materials are in the right place when needed. Once the task is complete, the workflow software makes sure that the individuals responsible for the next task are notified in advance and receive all the information they need to complete their stage of the process.

Other systems within CIM

As can be seen from the diagram, there are other sub-systems within the CIM including manufacturing cells.

Manufacturing cells

A fully automated **manufacturing cell** may be made up of just two CNC machines, i.e. lathe and milling machine, and a parts handling robot.

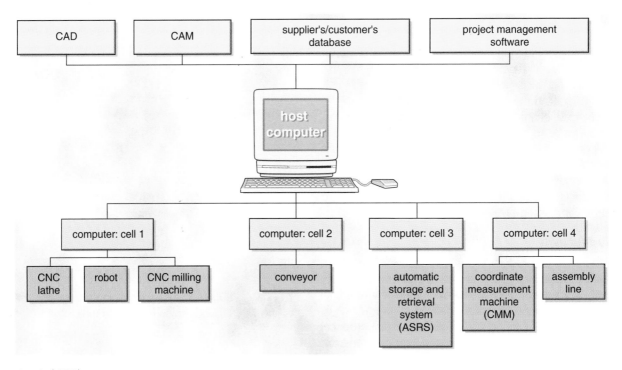

A typical CIM layout

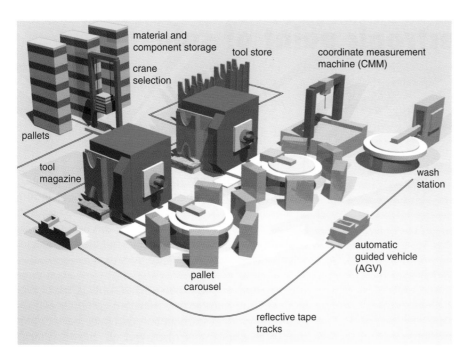

A fully automated manufacturing cell

For example, materials are transported from storage using an automatic guided vehicle (AGV) to the manufacturing cell. The robot takes the material (such as a steel bar) and places it in the chuck of the lathe. Once the bar has been turned to the programmed size and shape, the robot takes it out of the chuck and places it in a buffer store ready for the next operation. When needed, the robot takes the lathed steel bar from the buffer store and places it in the milling machine for further shaping. Once the component is completed, it is transported to the coordinate measurement machine (**CMM**) for quality control.

Automatic storage and retrieval systems (ASRS)

An ASRS system in a fully automated, computer controlled textile warehouse

On fully automated production lines, the host computer will control the transportation of materials and components to the required points on the assembly lines. This is achieved by using two main methods:

- a conveyor
- an automatic guided vehicle (AGV).

Conveyor

All stocks of components and materials are stored in pallets on large storage shelves. The ASRS system will select the correct component off the shelf by means of a crane, retrieve it and place it ready for the AGV to pick up.

Automatic guided vehicle

This is an unmanned vehicle that carries components automatically along a pre-programmed path. The simplest way of doing this is by using reflective tape fixed along the route of the vehicle. The AGV will 'read' this route using an on-board photo-sensor.

Coordinate measurement machine

A coordinate measurement machine (CMM) is a fully automated instrument, controlled by the host computer, for checking the measurements of manufactured components. It is a method of quality control because of its accuracy in measuring length, height, inside and outside diameters and other aspects such as flatness.

▪ Things to do ▪

1 What are the potential disadvantages of a CIM system?

2 What scale of productions is CIM ideal for?

Electronic point of sale

Aims

- To understand how the electronic point of sale system is used to gather product sales information.
- To understand the significance of the digits that make up a bar code.

Information is at the centre of any business and, if used properly, it ensures the business stays one step ahead of its competitors. By using electronic point of sale (**EPOS**) a business is able to supply and deliver its products and services faster by reducing the time between the placing of an order and the delivery of the product.

Electronic information is collected and recorded at the point of sale using a bar coding system on all products.

Bar codes

Each product can be electronically identified using its unique bar code. This universal product code (UPC) was set up by the Uniform Code Council (UCC) in the USA.

How a bar code works

When passed over a bar code reader or scanner, the bar code is read by a laser beam. Think of the checkout assistants at your local supermarket and the familiar 'beep' sound as your shopping is scanned. The laser or light beam passes over the bar code and reflects back on to a photoelectric cell. The bars are detected because they reflect less light than the back-

ground on which they are printed. Each product has its own unique number.

On the 13-digit bar code shown below, the first two numbers tell you where the product was made. The next five digits are the brand owner's number, and the next five numbers are specific to the type of product (given by the manufacturer). The final digit is the check digit, which confirms that the whole number has been scanned correctly. There are a number of variations on the 13-digit bar code, but they all work in a similar way.

The scanners at supermarket checkouts transmit the product code number to an instore computer which relays the product's description and price back to the checkout, where it is displayed electronically and printed on the till receipt.

The computer then deducts the item purchased from the stock list so that they can be re-ordered when necessary.

4 0	0 2 5 8 9	0 0 1 1 8	7
Origin of product	Brand owners number	Item number	Check digit

A standard 13-digit bar code

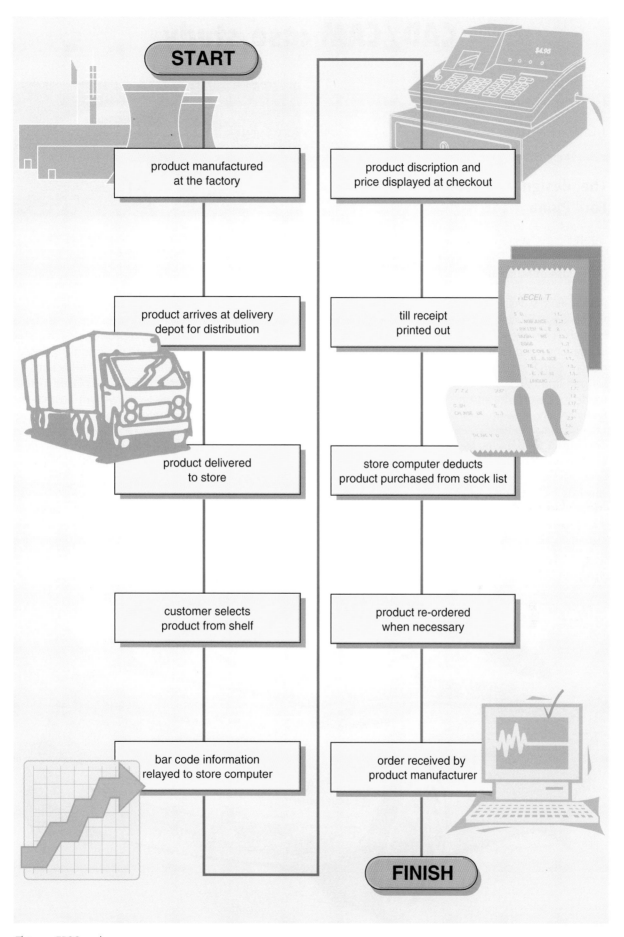

START

product manufactured
at the factory

product arrives at delivery
depot for distribution

product delivered
to store

customer selects
product from shelf

bar code information
relayed to store computer

product discription and
price displayed at checkout

till receipt
printed out

store computer deducts
product purchased from stock list

product re-ordered
when necessary

order received by
product manufacturer

FINISH

The way EPOS works

CAD/CAM case study

Aim

- To understand the use of CAD/CAM in developing products.

The design and development of the Ford Puma

The brief given to Ford's design team was to design a sports coupé with 'a sporty image and appearance'. The result was the first car ever to be designed solely by the use of computers.

Initially, Ford's design team brainstormed the look of the car by producing 50 or so concept sketches. The best six were chosen and developed into full-sized illustrations using Ford's 2D computer-aided industrial design (**CAID**) system. The system replaced the need for pen and paper and instead computer-generated renderings were drawn on screen using a graphics tablet and the powerful software.

Two of the six designs, codenamed 'blue' and 'red', were developed further by transferring them into a 3D CAID system. Then 3D wire frame **computer models** of both cars enabled the design team to refine the styling of the body shape and to calculate structural and engineering specifications. Computer modelling also indicated how the car would handle

The 'red' concept design, modelled above in the 3D CAID program, even has realistic lighting effects

even before any prototypes were produced and test driven.

Once the structure was complete, the wire frame models were given a 'skin' using computer-generated materials. This produced a realistic looking image even down to lighting effects and the reflections on the metal bodywork. Each car could be placed in 'real-life' surroundings such as a showroom so that the design team could evaluate its appeal. Animations were also produced blending computer renderings of the Puma with actual video footage in order to create a 'virtual reality' of the car driving down a road.

Using the exact information generated from the 3D CAID system, Ford was then able to manufacture precise full-sized clay models. Ford invested in two five-axis milling machines which could cut both sides of the

After brainstorming, the best ideas were drawn in the 2D CAID program. The 'blue' concept sketch, shown above, was the one finally approved

Milling machines can make small or full-sized clay models

model at once. The five-axis milling machines were able to change their orientation in five directions as they cut and had an accuracy of +/–0.075 mm due to them being directly controlled from the CAID software.

The combination of **'virtual' design** and clay models decreased the number of actual prototype cars having to be built to carry out important road tests. In addition to the several prototypes produced, computer modelling also allowed vital calculations to be made in simulated crash tests.

From the initial concept sketches to final approval, the Ford Puma took just 20 weeks to design and develop. To put this into perspective, it would usually take a team of 12 people about 12 weeks to develop a concept sketch into a clay model. Now, with the use of powerful computers and software, it can take just one person less than three weeks!

Ford's use of **CAD/CAM** dramatically speeded up the design and development process because:

- early 2D renderings were quickly generated and easily corrected on a CAD system
- 3D wire frame models enabled computer modelling to work out vital calculations
- 3D renderings and animation created a 'virtual' image of the proposed design

- 3D models provided accurate data for the production of full-sized clay models using CNC milling machines.

Future technology

Ford is currently developing 'holographic imaging' in order to speed up even further its design process and to improve quality. This new technology will be used by Ford to create full-scale virtual digital models of prototype vehicles. It will enable designers to manipulate a 3D holographic image of the car using their hands to modify the design.

Ford will also be able to project a holographic image of the car's instrument panel to view it from behind the wheel. This would enable customer feedback months earlier than is possible now by eliminating the need to make actual prototype cars. The project will dramatically reduce the time needed to develop a product, so saving costs and improving quality by use of virtual testing.

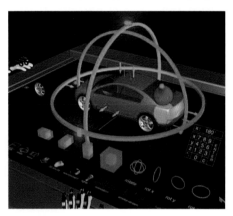

Computer-generated renderings speed up the design and development process

A realistic animation was created by mixing a design rendering with video footage

■ **Things to do** ■

1 The automotive industry has benefitted from CAD/CAM, but how could this be used in the packaging industry?

2 Visit your local DIY superstore and investigate how kitchen planning design has benefitted from CAD.

Director of Design Claude Lobo with the finished Ford Puma which took just 20 weeks to design and develop

ICT within design and development

- To understand how computers are used for creating 'virtual products'.
- To understand how computers are used for collecting, analysing and exchanging data.

Virtual products

Virtual reality is a relatively new technology using 3D CAD systems. These systems can create virtual reality environments and products that allow the viewer to choose where to look and travel. A virtual product can be viewed from any angle at any distance, using built-in camera angles, and with any moving parts in operation. The main advantage of virtual reality products is that they can be viewed on screen before they are actually built, so saving the time and costs of producing actual prototypes.

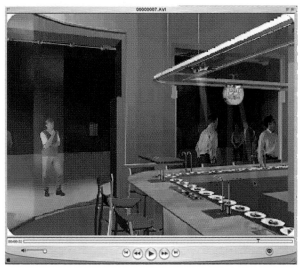

Using virtual reality architectural modelling, it is possible to take a tour of a new Yo!Sushi restaurant before it is even built

Virtual reality modelling

Virtual reality modelling is particularly useful for architectural design where a full-size prototype is not possible. Buildings can be modelled on screen to show the internal and external organization and layout. This technique can also show how a building may look in different materials or colours, and even at different times of the day as the position of the sun changes. By using various camera angles built into the computer program, it is possible for the architect to take a 'virtual tour' of the building before it is even built.

Switzerland to Sydney – from International Olympic Committee Headquarters in Lausann the IOC president listens to the Sydney Organising Committee for the Olympic Games president report by video conference

Electronic data exchange

As the majority of modern media (magazines, books, etc.) is computer generated, files can be sent digitally from the designer to the printing company. This eliminates the time and cost of physically sending work to a printer. Large files can be sent at high speeds using ISDN (**Integrated Services Digital Network**) without any loss of quality. For example, in publishing it is now possible for a book to be designed in the UK and printed in Hong Kong using computer-to-plate technology.

Video conferencing

Video conferencing has opened up a completely new way of communicating with people in other parts of the country or across the world. A company may use video conferencing to communicate simultaneously with several people in different locations. Each location has its own digital video camera into which people talk while, at the same time, seeing the other locations on their monitors. Therefore, important business meetings with company executives, designers and manufacturers are possible using a three-way communication link. This eliminates the time and cost involved in having to travel to meetings.

Remote manufacturing

Another important aspect of video conferencing technology is the development of **remote manufacturing** of components. Pioneered by Denford Ltd, the computer-aided design of a component by a student in a school can be manufactured miles away at Denford's Remote Manufacturing Centre using computer-aided manufacturing processes.

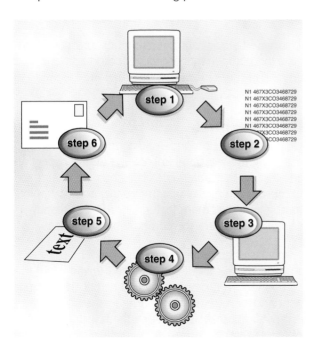

Remote manufacturing

The process involves the design of the component using CAD/CAM software and then live discussion between technicians at Denford and the student concerning the project. The student's design is then sent digitally to Denford, where any adjustments are made prior to making. The design is manufactured using a CNC machine while the student views via a video camera. The finished component is checked for quality and posted back to the school. Remote manufacturing is particularly useful in schools where the necessary equipment is not available to produce a batch of identical components.

Computer-aided market analysis (CAMA)

Market research is used by businesses to gather data about consumers, the needs of the market and to find out how effective their marketing strategies are. Market analysis focuses on the analysis of data along with existing research data to predict the future of a particular market (or market trends).

The collection of the data can be done 'in-house' by a company or by a specialist market research agency (Have you been asked to complete a questionnaire on the street or on your doorstep?). The vast amounts of data collected are entered into a computerized database. The database can then be used to help the company:

- to make sound marketing and planning decisions
- to calculate the demand for its products or services
- to identify markets and find out where potential customers are shopping
- to target specific groups of customers
- to launch new products more effectively by using more focused marketing strategies, for example regional launches instead of national.

In modern business, a company needs access to up-to-date research, in-depth product and market analysis and industry-specific expertise to make the best marketing decisions. The use of advanced computer-based marketing tools and expertise will help to achieve this goal.

ICT within manufacturing

Aims

- To understand how the use of computers can enable faster and more flexible manufacturing.
- To understand the use of computers for stock and quality control within manufacturing industries.

Many aspects of modern manufacturing are reliant upon computers and the use of information and communications technology (**ICT**). The use of computers ranges from the control of manufacture using computer numerically controlled (**CNC**) machines to stock control and the flow of information throughout a company. Information and communications systems have greatly increased the productivity and competitiveness of manufacturing companies.

Computer numerical control

This is part of the process of computer-aided manufacture (**CAM**). CNC literally means the control of machines using numbers or digital information. CNC machines were first developed in the 1950s and remain very popular within manufacturing for drilling, turning using lathes and milling using vertical milling machines.

Older CNC machines required the operator to program in very complex codes in order to make a component. However, it is now possible simply to draw a profile of the component on screen and the sophisticated software will generate all of the complex codes without the use of programming. It is also possible to 'virtually manufacture' the component on screen before actually making it using the CNC machine.

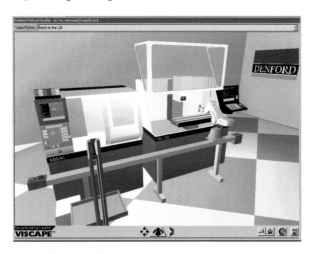

Denford's virtual milling centre

CNC machinery has developed into versatile manufacturing centres, and fully automated manufacturing cells are being used to provide maximum flexibility. Manufacturing centres are staffed by a small number of skilled, versatile technicians to make sure that these high cost machines operate at maximum efficiency. Centres form part of the integrated network of computers that make up computer-integrated manufacture (**CIM**).

Advantages of CNC machines

- Products are made accurately and at speed.
- They are flexible, i.e. they can be used in both batch and mass-production systems.
- They are reliable, i.e. they can be used continuously and in situations that are hazardous to humans.
- They are economical to operate.

Disadvantages of CNC machines

- They are expensive to buy and set up.
- They are complex and can break down bringing the whole of a production line to a standstill.
- They require highly skilled and expensive staff.

Just-in-time

Just-in-time (JIT) is a method of production management used to meet customers' orders with the minimum of delay, to the required level of quality and in the right quantity. JIT uses computer-aided management of stock in which materials and components are delivered from outside suppliers just before they are needed at each stage of the manufacturing process.

Automated stock control systems ensure that materials and components are well stocked and available on demand when and where they are needed, so reducing waste.

Quality control

Quality control is concerned with monitoring and achieving agreed standards as a result of inspection and testing. Computer-aided methods are often used to monitor the quality of components at critical points throughout the manufacturing process. Testing is carried out to check the components' tolerance.

Computer-aided testing usually involves a machine with a probe which makes contact with the surface of the component. The probe is moved across the component and its measurements are taken and fed into a computer for analysis. Other methods include the use of optical sensors as used in the printing industry.

Colour bars

Colour bars provide vital information about the performance of both the printing press and the inks being used. They contain a whole range of tests, some are visual checks made by people but others can be done electronically using a **densitometer**. This instrument monitors the thickness or density of the ink printed on the colour bar to ensure that it is of a consistent quality throughout the print run.

Advantages of using computer-aided quality control

- It provides ongoing quality control throughout manufacture.
- Different types of inspection processes can take place by using different probes or sensors, for example contact, non-contact and vision systems.
- Data from the probes or sensors is directly inputted into a computer system and reports can be generated.
- Errors are identified, analysed and corrected early, which reduces waste and therefore costs.

■ Things to do ■

1 Collect examples of colour bars from packaging or magazines.

2 Why do manufacturers invest vast amounts of money in computer aided quality control rather than relying upon human inspectors?

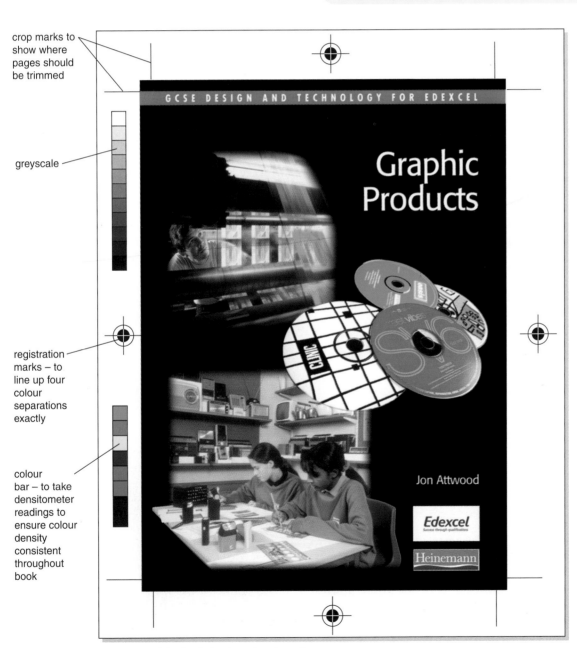

crop marks to show where pages should be trimmed

greyscale

registration marks – to line up four colour separations exactly

colour bar – to take densitometer readings to ensure colour density consistent throughout book

Front cover of this book showing printer's marks

Practice examination questions

1 Describe the process of producing custom made vinyl stickers in small scale production using CAD/CAM software. **(6 marks)**

2 CAD systems are often used in fitted kitchen showrooms to produce plans for customers. Explain two advantages of using a CAD system for this purpose. **(4 marks)**

3 Explain three advantages of using CAD when developing a mass-produced product such as the one in the photo below. **(6 marks)**

4 Describe two uses of ICT as a data handling system for a large company using mass production processes. **(4 marks)**

5 Explain how batch production processes benefit:
a) the manufacturer
b) the consumer **(4 marks)**

6 Name four pieces of information stored on a products bar code. **(4 marks)**

7 Explain how Electronic Point of Sale (EPOS) is used to monitor stock levels in a supermarket. **(6 marks)**

8 Discuss the benefits of advertising virtual products on the Internet. **(4 marks)**

9 Computer Integrated Manufacture (CIM) is used in modern methods of production. Describe how CIM increases efficiency in terms of:
a) materials
b) time **(6 marks)**

A full size elevation view of a Chrysler car, generated by a CAD package. CAD is an important tool in industrial design

Section D:
Design and market influence

Colourful utensils made of a biodegradable, compostable plastic from corn

Consumer issues

British Standards Institute

The British Standards Institute (BSI) is an independent organization concerned with publishing technical documents called 'British Standards'. The BSI also works closely with the European Committee for Standardization known as CEN (**Comité Européen de Normalisation**) in providing relevant standards throughout Europe. By testing, inspecting and issuing certificates to products and companies, these organizations help to make sure that what the consumer purchases will do the job for which it is required.

The BSI Kitemark Comité Européen Mark

When designing a product, it is important that you are aware of the relevant British Standard. Most libraries will hold a reference copy of the BSI catalogue and many have copies of the actual standards on microfiche. There are literally thousands of specific standards, for example BS 1133 Sections 1 to 3: 1989 concern a specific code for the selection of packaging materials.

Testing by the BSI informs the government and manufacturers of the standard for quality and safety in most products. The need for tamper-proof and child-proof packaging is extremely important for a number of products. For example, the following is part of a series of tests required to be carried out for 'child-proof' medicine containers such as paracetamol bottles if they are to be awarded BS 5321. (The sample is of 200 children aged 42–51 months.)

- At least 85 per cent of the test panel of children shall be unable to open the containers prior to the demonstration.
- At least 80 per cent of the same children shall still be unable to open the containers after the demonstration.
- At least 90 per cent of a panel of adults shall be able to open and properly re-close the containers by following written instructions, but without having received a demonstration.

Food package labelling

The Food Safety Act 1990 is concerned with consumer protection. It states that any food available for retail sale should have a printed wrapper or container which:

- does not 'falsely describe the food', or
- is not 'likely to mislead as to the nature or substance or quality of the food'.

If any packaging fails to meet this, then the manufacturer is guilty of an offence and is liable to prosecution.

Food Labelling Regulations 1984

The Food Labelling Regulations 1984 set out the labelling requirements for all types of food. The following information must be clearly visible on the packaging, or a label attached to the packaging:

- The name of the food must describe the product and not be misleading, for example 'chocolate flavoured' when it does not contain real chocolate.

What food package labelling should not be

added nutritional claims

* best before date

* storage conditions

company name

nutrition/information

* ingredients (in descending order of weight)

promotional offer

* weight (estimate)

* name or description of the food

added nutritional claims

* name and address of manufacturer, packer or seller

Food Labelling Regulations

* required by law (1984 Food labelling Regulations)

- A list of ingredients (greatest first), including all of the additives used named as E numbers.
- A best-before date (and sell-by date for perishable food).
- Any special storage conditions or conditions of use.
- The name and address of the manufacturer. Packagers or seller within the European Union must be shown.
- Details of the place of origin of the food.
- Appropriate instructions for preparation if applicable.
- Quantity or weight. The 'e' after the weight is a European weights and measures system that means that it is an average and individual packs may vary slightly in weight.

■ Hints and tips ■

When designing your own packaging, make sure that the label clearly includes all of the necessary legal labelling requirements.

■ Things to do ■

1 Why is the test for a child-proof paracetamol bottle appropriate?

2 Analyse a piece of packaging by labelling the information required by law.

Electronic communications

Aim

- To understand the possibilities of new technologies.

Global networks (the Internet)

One of the major developments of the late twentieth century – the **Internet** – is providing an excellent means of accessing a wealth of information and entertainment and a new market place for companies to do business. Anyone with access to a computer and modem can 'surf' the world wide web by using a web browser (for example Netscape Navigator or Internet Explorer) and an **Internet service provider** (ISP) such as BT or AOL. It is even possible to access the Internet from certain types of mobile phone enabling people to work on the move.

E-commerce

The term electronic commerce or e-commerce describes the many ways in which a company or business can advertise and sell its products or services using the world wide web. The main advantages for business are:

- a global market place
- product information, catalogues or services displayed on screen
- easy to update information
- online shopping to sell products (increased sales)
- improved service to customers.

E-mail

Another important aspect of the Internet is its ability to allow people from all around the world to communicate with one another using electronic mail or e-mail. Nowadays, it is relatively simple to set up your own e-mail account using an ISP. ISPs such as Hotmail and Yahoo allow you to access your e-mail from any Internet-ready computer in any country in the world!

Disadvantages of the Internet and e-mail

- Access to inappropriate material or information, therefore the need for filtering of information for particular age groups.
- 'Junk' e-mail and the spread of computer viruses.
- Questions of copyright.
- Not accessible to everyone.

Electronic data exchange

As the majority of modern media (magazines, books, etc.) is computer generated, files can be sent digitally from the designer to the printing company. This eliminates the time and cost of physically sending work to a printer. Large files can be sent at high speeds using ISDN (Integrated Services Digital Network) without any loss of quality. It is now possible for a designer working in the USA to send digital artwork for printing in the UK.

Just some of the many Internet service providers

Modern website graphics aimed at the youth market

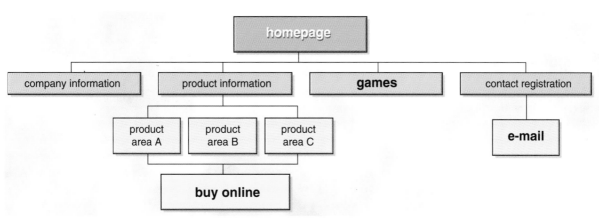

A website navigation tree

Website design and graphics

The Internet has opened up a brand new sphere of graphic design – website design. Most companies now have an Internet site and graphic designers have been commissioned to design their websites in order to advertise and entertain potential customers. Websites come in all styles and sizes according to the company or product they represent. For example, a website for a cola drink will be fun with funky graphics and typefaces in order to appeal to the youth market, whereas a building society may use more upmarket images in order to inspire confidence.

A website can contain many different pages, just like a book. It is possible to scroll down any single page by using the mouse and tool bar to read text. More importantly, the use of hidden buttons contained in text or graphics can link directly to other pages just by a click of a mouse button. Links from a company's homepage can take you to different areas of its website such as company profiles and history to specific product information and order forms. Links can also be made to areas outside its own website that can put you in touch with other related websites.

■ Things to do ■

The Internet is available in many schools and is an invaluable tool for carrying out design and technology activities.

1 Use your school's web browser to type in a key word relevant to your project in order to obtain research information.

2 Study a website that you particularly like. Analyse its style and target audience, then draw a navigation tree for the site.

Smart materials

Aims

- To understand the new technologies of smart materials.
- To understand the smart materials used in graphic products.

Smart or modern materials have been developed through the invention of new or improved technologies, for example as a result of manufactured materials or human intervention, in other words, not naturally occurring changes.

Smart materials respond to differences in temperature or light and change in some way as a result. They are called smart because they sense conditions in their environment and respond to those conditions. Smart materials appear to 'think' and some have a 'memory' as they revert back to their original state. The term smart can be quite unclear as in some cases it is difficult to distinguish between modern and smart materials.

Polymorph (polycapralactone)

Polymorph is a new plastic polymer that is classed as a modern material rather than a smart material. It has a very low melting point (62°C) and can be melted in hot water. Once heated, it becomes a soft and pliable material with the moulding properties of plasticine. Therefore, it can be used to produce rapid prototypes for graphic products such as hand-held devices to test ergonomics. Once cooled, polymorph has the advantage of hardening like a plastic whereas plasticine remains soft.

Thermochromic film

This is a type of liquid crystal display (LCD) commonly found in the screens of calculators and laptop computers. Thermochromic film is perhaps best illustrated in the test panel in many batteries now available. It is a smart material that responds to changes in heat from electrical current by changing colour.

The test panel is made of thermochromic liquid crystal which is put into minute capsules by a process called microcapsulation. These capsules are made into an ink that can be printed on to plastic or paper. In battery test panels, a good battery causes enough electrical current to pass through the battery's plastic jacket to heat up the ink. The heat then causes the panel to change colour, usually a bright yellow. If the battery is running low on charge, then there is less heat generated and the test panel reads a shorter yellow bar to alert the user.

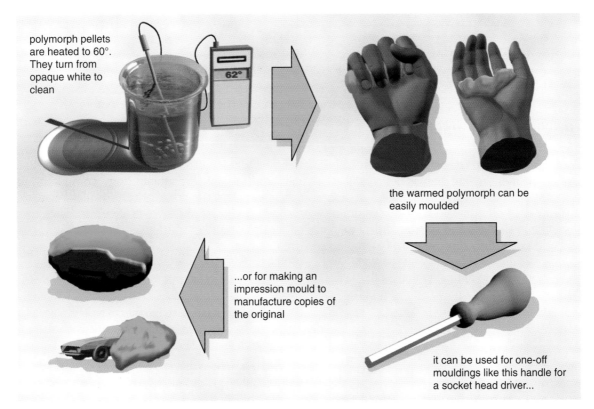

polymorph pellets are heated to 60°. They turn from opaque white to clean

62°

the warmed polymorph can be easily moulded

...or for making an impression mould to manufacture copies of the original

it can be used for one-off mouldings like this handle for a socket head driver...

Polymorph

The Technology Enhancement Programme (TEP) provides a simple kit for producing a thermocolour display that could be used in your projects.

The kit comprises a self-adhesive plastic film which is then overprinted with thermochromic liquid crystal ink and a length of resistance wire. Using the wire, you can form a logo, symbol or message which when connected to a battery causes the display to light up in the shape of your design.

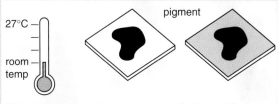

If black thermochromic pigment is applied to a white surface, the surface turns from black to white at the change-over temperature.

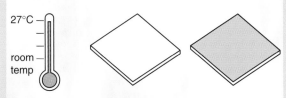

If the pigment is applied to something orange, the surface colour changes from black to orange at 27°C.

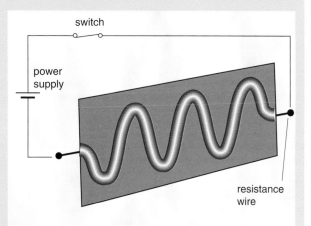

A visual display idea based on the use of resistance wire behind a sheet of purple coloured paper covered with black thermochromic pigment. Where the wire heats up the pigment, the purple starts to show through.

The TEP's thermocolour display kit

Battery test panel using thermochromic film

Other applications of thermochromic film can be seen in thermometers and warning patches on computer chips to show when they are getting too hot.

Advanced liquid crystal displays

Scientists are currently working on smart materials to increase the effectiveness of liquid crystal displays. For example, currently the screens on laptop computers have three main disadvantages:

- LCD screens are either limited to shades of grey or dim, dingy colours.
- More advanced versions, capable of showing bright, fast changing shades of colour are expensive.
- All LCDs use a lot of power in order to operate.

With the development of cholesteric liquid crystals it will be possible to design smaller pixels on the screen resulting in higher resolution screens with much brighter colours. The crystals will be more stable which will eliminate the screen flicker that makes laptop screens wearisome to read when used for long periods. Perhaps the major advantages will be that they will use less power, therefore running for longer on batteries (more than ten times longer) and they will be considerably less expensive to buy.

This is only the beginning for smart materials – they will soon be in everything from computers to concrete bridges! Sensors made from smart materials could be embedded inside car tyres and concrete bridges to monitor their wear, and even biological smart materials could be used inside the human body as artificial muscles.

■ **Things to do** ■

1 Try to find out more about smart materials in use today, for example shape memory alloys (SMA).

2 Use the TEP's thermocolour display kit in your design activities to produce colour displays for models or signage.

Design and market influence

The impact of CAD/CAM

Aims

- To understand that CAD/CAM has enabled products to be made in quantity and cheaply.
- To understand that CAD/CAM processes have social implications for the workforce.

The need for manufacturing industries to develop products that are competitive in modern global markets is vital for their own survival and the prosperity of their workforce and other businesses in their local community. Eventually, most products can be designed better or updated because of advances in technology or produced more economically by improvements in production methods. The use of computer-aided manufacturing (**CAM**) systems has helped to make components and products rapidly, accurately and less expensive.

Designers can use computer-aided design (**CAD**) to create, develop and communicate product design information. Computer-aided manufacturing technology is developing rapidly to translate CAD information into manufactured products. The combination of **CAD/CAM** using computer-integrated manufacture (**CIM**) is providing a system of designing, planning, manufacturing and monitoring the production of products on a large scale.

Social implications of CAD/CAM

CAD/CAM technologies are directly affecting patterns of employment in manufacturing right across the globe. The numbers of people involved in manufacturing are declining as automated systems take over. For those who are left, their jobs have changed significantly.

Effects of CAD

Designing used to be carried out in large drawing offices with lots of people working on the different components of the same job. Skilled technicians and draughtsmen hand drew all of the technical information required to manufacture the product.

Now, a small team of highly skilled computer **designers** can perform the same tasks more efficiently and in less time. Many draughtsmen have had to retrain to become computer literate and familiar with industry standard software packages. Those unwilling or unable to retrain either have less opportunity of work or become redundant.

Car designers work on design drawings at the Ghia automobile company in Italy in 1956

Skilled computer designers have taken over from skilled draughtsmen. Here, a car designer creates a computer aided design of a car on a computer display screen

Effects of CAM

Traditionally, workers in manufacturing industries have been highly skilled machine operators. School-leavers followed apprenticeship courses within the employer's factory in order to train them in the operation of a particular piece of machinery. Often young people would serve their apprenticeship and work for the same employer until they retired.

Operators of CAM machinery now have to be trained to operate different machines and carry out many processing tasks within their manufacturing cell or work centre. Gone are the days of narrow specialization when, for instance, lathe operatives were not allowed to set up and use another person's machine in the factory. They now have to work as part of a team offering help to others if required.

The modern workforce

The modern workforce has had to become skilled, trained and above all flexible. There are severe shortages of multi-skilled and multi-function operators in

the UK at present as a direct result of modern manufacturing processes.

A significant number of people working in the manufacturing industries will be unskilled and low-paid machine operatives, assemblers or packers. With developments in automated machinery and the use of robots, there will be even less human involvement in the manufacturing and production processes of the future. This can create emotional and other problems at work. When CNC machines are controlled directly from a central system, some jobs consist of nothing more than 'machine minding'. This, in turn, can lead to poor job satisfaction and reduced productivity. There is now an important need for companies to devise systems and develop schemes to make their employees feel valued.

■ **Things to do** ■

1 Do the advantages of CAD/CAM outweigh the social implications?

2 Discuss the social implications of workers working in an automated factory.

Moral issues

The rise of consumerism

We live in a consumer society where we are constantly bombarded with images of brand new products that we are told we 'need'. There are vast ranges of products satisfying identical needs and all of these may be purchased with relative ease. Products are now designed for a mass market creating multi-national companies selling and trading in global markets.

Mass marketing

Mass marketing involves developing products with widespread appeal and promoting them to all types of customer. The ultimate aim of mass marketing is to develop a global name such as Nike or Coca-Cola. Mass marketing, however, relies on reducing costs and selling large amounts of products with a relatively small profit margin.

In order to produce higher profits, companies have to operate in more than one country and become multi-nationals selling in a global marketplace. This requires heavy investment and lower manufacturing costs resulting in many companies having their products made in developing countries where labour and over-heads are considerably less expensive.

Different cultures

To be successful in this global market place, a company has to have a product which appeals to people from different cultures. Some products have to renamed in certain countries in order for them to be internationally recognized. For example, the Snickers bar used to be called Marathon in the UK, before Mars renamed it because it was already known as Snickers in many other countries, including the USA. Sometimes a product is sold under another name in different countries as the original name might cause offence.

The product lifecycle

The product lifecycle shows the sales of a product over time.

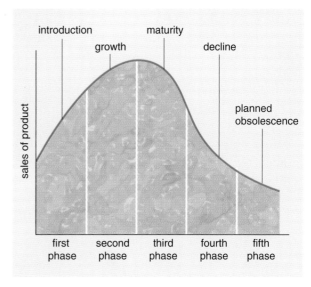

The product lifecycle

Introduction

This is the initial launch of a new product to satisfy a potential need in the market. During this phase sales are usually low due to lack of reputation, and high costs involved in product development mean that the product is highly priced. Many products fail at this stage.

Growth

If the introduction is successful, then sales will grow. Growth is characterized by a sustained rapid increase in sales.

Maturity

Eventually sales of the product will level out, and there will be little or no further increase. This may be due to many factors including market saturation (where the product has reached all parts of the market) and the fact that a potential need has been satisfied.

Decline

Sales of the product decrease below the level of sales during maturity. All initial investment costs for product development should be covered by this stage.

Planned product obsolescence

In a society that is constantly changing due to techno-logical advances some products can quickly become out of date or obsolete. Companies are all too aware of this and may launch a product knowing that it will have a short product lifecycle.

During the 1960s and 1970s there were famous cases involving some American car manufacturers who made constant styling changes to their range and even developed cars that would rust easily. As a result, the general public tended to upgrade their cars more often than usual. Another American company carried out market research as to why the public had to buy a new potato peeler. They found that the average person lost their peeler by accidentally throwing it away along with the potato peelings. Therefore, a new model of potato peeler was launched with a murky coloured brown handle to camouflage itself in with the potato peelings.

Nowadays, with the impact of the home computer market, companies will constantly upgrade the specification of their personal computers (PCs). The speed of this has increased so that once you buy a PC there is always a newer and more powerful one for sale almost immediately. Other products such as point-of-sale displays for shops will be made out of materials that will not stand wear and tear. The company may simply want the display to last for a certain amount of time during a marketing campaign. Once the campaign is over there is no longer a need for the display and it is either destroyed or sent back to the company's warehouse for storage.

Changing fashions

Changing fashions, especially in youth culture, have a very important effect on the development and life-cycle of products. As one product becomes popular and sales increase, then another becomes unpopular and sales decrease. Manufacturers are only too aware of this fact and they will update or change the brand identity of the product in order to modernize it.

■ Things to do ■

1 Discuss the moral issues of large multi-national companies using the Far East as a source of inexpensive labour to produce their products.

2 In terms of planned product obsolescence, discuss the reasons why most Premiership football clubs change their strip every season and introduce a number of 'away' strips.

3 Make a list of products that were popular last year. Compare this list with this year's fashionable items to own or be 'seen wearing'.

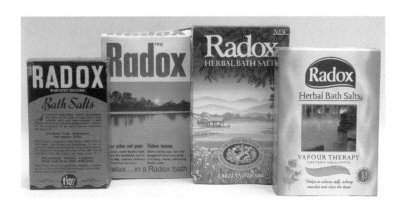

Changing the brand identity over time

Environmental issues

Aim

- To understand that environmental issues affect the design, manufacture and disposal of products.

Environmental issues are gaining much attention in a technological society with major manufacturing industries and mass consumerism. Designers, manufacturers, retailers, governments and consumers all have an important role to play in providing 'green' solutions to design problems and sustaining the world's natural resources.

Sustainable technology

Sustainable technology involves:
- recovery of materials
- recycling materials
- using recycled materials.

Most materials can be recycled, which saves energy and can help reduce pollution but, most importantly, means that less raw materials have to be extracted and processed. The initial problem, however, is how to recover sufficient materials to enable large-scale recycling. It is now a common sight for large supermarkets to employ recycling schemes and collection points for a wide range of materials including glass, paper, metal cans and textiles. Once materials are recycled, they must be made into products that the consumer wants to buy in order for there to be a sufficient market for them.

Designing for sustainability is very important for designers. By designing products with longer lives there is less need to buy new ones which will in turn use more energy and materials in their manufacture, for example upgrading computers instead of buying new ones.

Pollution

Every product we buy requires energy in its manufacture. Most of our energy is generated from the burning of fossil fuels such as coal, gas and oil. These all produce harmful gases when burnt and have increased concerns about global warming and acid rain. Petrol used for powering vehicles for the transportation of products and cars for domestic use also produces harmful gases, especially lead. The growth of public concern for atmospheric lead pollution has led to the increased use of lead-free petrol and catalytic converters in vehicles and the government increasing the road taxes for high-powered cars.

Landfill

Most of our household waste ends up at a landfill site where it is buried. These waste materials often take years to decompose and require strict control measures by waste management companies to reduce pollution such as smells, run-off of pollutants into local water supplies and visual pollution from large-scale tipping of waste. The development of biodegradable plastics such as Biopol that are made from the fermentation of food waste, means that such materials will decompose more rapidly when buried on landfill sites.

Most of our household waste ends up at a landfill site

An extrusion machine producing polythene sheet tubing – plastic tubing is blown out of its former shape by hot air, which shapes and dries the plastic before it is folded and rolled. Recycled polythene is used in the building industry

PVC

Plastics have had a poor environmental image, probably due to them being made from crude oil and as visible forms of litter in our streets. In particular, PVC has been singled out as a major cause of pollution because chlorine is used in its manufacture and it has the potential to produce hydrochloric acid and dioxins when incinerated (burnt). Some countries have banned the use of PVC as a result. For example, some European countries do not allow the use of PVC shrink-wrap sleeves in packaging and instead use PET which is seen as less damaging to the environment.

However, PVC is probably less harmful to the environment than alternative materials. It requires less energy to manufacture than other plastics because more than half of its composition is chlorine. The chlorine itself comes from salt – a plentiful resource. In addition, harmful dioxins and hydrochloric acid can be successfully removed by scrubbers when incinerated.

Conservation of resources

One of the first rules for all designers should be to minimize the amount of materials used in a product whenever possible. The benefits of this are far-reaching – from the conservation of resources, through the reduction of energy and pollution involved in manufacture to the reduction of waste materials for disposal.

One approach to conserving resources is to reduce the amount of materials used in packaging. Biscuits and chocolate, for example, often have several layers of packaging which, of course, is simply thrown away after use. Designers must look for ways of minimizing the use of packaging materials, either by reducing the number of layers or by reducing the thickness of the materials. For example, the walls of an aluminium drinks can are just 0.1mm – the thickness of a good quality paper.

Another approach may involve companies employing refill schemes to promote the re-use of their packaging. For example, some fabric conditioners available at supermarkets are bought in plastic bottles. When the customer has used the contents, a smaller 'soft pack' using a plastic laminate can be used to refill the bottle. This type of scheme, however relies heavily upon the consumer to participate so it must be financially rewarding for them to cooperate.

Paper and board are used routinely and millions of tonnes are discarded every year in the UK alone. Although wood is renewable (trees can be replanted), it is no reason to be wasteful. The paper making industry has therefore increased its production of recycled paper and board.

The largest users of recovered paper fibres for use in recycled materials are corrugated board, newsprint and toilet tissue manufacturers. There is an increasing number of recycled paper products available including writing paper and fast-food packaging. Many people believe that the use of paper bags instead of plastic bags is better for the environment. This is not strictly true:
- More energy is needed to produce paper bags.
- The manufacture of paper bags produces more pollution than plastic bags.
- Plastic bags are more likely to be re-used.

■ Things to do ■

1 Discuss how the following can promote environmental issues:

a the designer

b the manufacturer

c the retailer

d the government

e the consumer.

2 Examine the price difference and image of products that claim to be environmentally friendly.

Japanese design

Aims

- To understand the thinking behind Japanese design.
- To understand the influence of Japanese design upon the rest of the world.

One hundred years ago, Europe was the centre of the world's wealth and power, but today the situation is very different. The emergence of east Asia, especially Japan, marks a shift in the balance of world economic power. Japan has transformed itself into a financial and technological superpower exporting both products and culture all over the world.

Think of the major electrical product manufacturers and Sony, Nintendo, Panasonic and Sharp may spring to mind – all Japanese. These companies produce technological gadgets that we didn't think we needed until they marketed them. For example, until Sony developed the now famous Sony Walkman, we didn't listen to music on headphones walking down the street.

Japanese companies will tend to try out thousands of new products on the public every year, a few of which may be successful. This is completely different to European companies which will laboriously plan, test and be selective in the release of new products. This has enabled Japanese companies to be both creative and innovative and stay one step ahead in total sales of products.

A wide range of design in Japan takes a lot from Buddhism, which is a widespread religion there. Japanese design is renowned for its:

- simplicity
- compactness
- precision in detail.

Buddhism strives to simplify life and asks you not to think of yourself as an individual but part of a greater natural world. Therefore, a lot of Japanese design takes its influence from organic shapes rather than man-made machinery. For example, the growing of Bonsai trees is popular in Japan. A Bonsai tree is an exact miniature of a full-grown tree, still in complete proportion and said to keep all the 'power' of the full-grown tree. This idea has been transferred to electronic products where **miniaturization**, along with advances in technology, has produced some extremely innovative products.

Miniaturization through technical advances

Graphic design

Modern graphic design in England has been heavily influenced by Japanese logos and characters. This is most notably taken from Japanese *Manga* cartoons depicting violent science-fiction action featuring high tech machines, futuristic cities and stylized characters.

Such influences are found on clothing logos, artwork for albums and computer games.

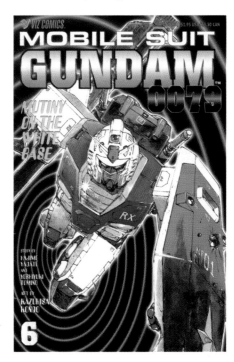

A typical Japanese Manga

The work of Sheffield-based The Designers Republic has explored the use of Japanese lettering in its futuristic artwork and design

Cute (Kawaii)

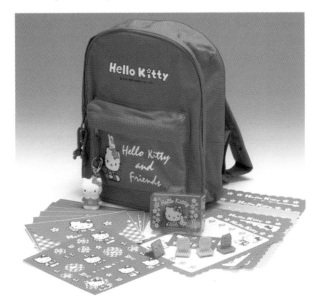

Hello Kitty merchandise

Some Japanese products are influenced by cute characters such as Hello Kitty. Hello Kitty was created by Shintaro Tsuji in 1974 as a character to use on greetings cards. Since then, the Hello Kitty character has been applied to a wide variety of merchandise from girl's T-shirts, hair grips, small purses and handbags to notepads, pencils and miniature bubble-gum dispensers. In Japan, where Hello Kitty is as part of the national culture as Mickey Mouse is in the USA, you can buy TVs, toasters, mobile phones and even wedding dresses with the Hello Kitty character.

▪ Things to do ▪

1 Discuss, using your own examples, the influence of Japanese design upon your quality of life, for example the miniaturization of products.

2 Compile a list of Japanese manufacturing companies from a trip down your local high street. Compare this list with European companies making similar products.

Analyse and evaluate products and processes

Aims

- To understand the importance of analysing existing products.
- To understand the need for quality of design and quality of manufacture.

It is important that we **analyse** existing graphic products in order to give us an insight into the work of professional designers and how they have satisfied a design brief. We need to do this to gain a greater understanding of commercial design and industrial manufacturing processes which will in turn inform and influence our own design-and-make activities.

Product design specification

A design **specification** sets out the essential criteria that the intended product should achieve. It is essentially a checklist of bullet points against which the final product can be tested and evaluated. Every product would have had a design specification and we can use that specification to judge the quality and effectiveness of a graphic product.

Important questions to ask are:

- What is the aim of the product? (Its purpose)
- What should the product do to achieve its purpose? (Its function)
- What is the market for the product? (The end user and target market)
- How does it look? (Its aesthetics)

- How is it made? (Materials, processes and scale of production used)
- Is it safe to use? (Quality issues and safety standards)

We can try to find the answers to these questions by using several methods of analysis including:

- focused case studies
- disassembly of a product(s)
- comparison of similar products.

Quality of design

Quality is of great importance in the design of a product since no product will sell very well if it is of poor quality. When we buy a new product we may have two things in mind:

- How good does the product look? (Its aesthetics)
- How well does the product work? (Its function)

Our initial reaction to a product will naturally be its looks. Manufacturers realize that a product must look good in order to gain the attention of the consumer. Market trends such as product styling, colour and fashion have to be taken into account so that consumers purchase a product that expresses their way of life. For example, Apple's i-Mac computer changed people's views about what a household PC should look like. Its styling was so different that people wanted them as a fashion statement and Apple's new i-Mac, launched in January 2002 has taken design ideas further.

Successful product styling – the evolution of computer hardware design

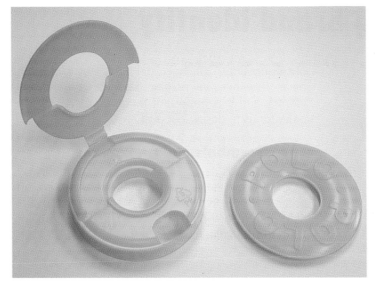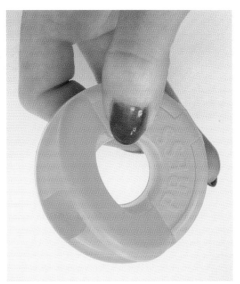

Example of injection moulding – the mini POLO dispenser

On the other hand, a product is of no use if it simply looks good. The product must work well in order to have quality of design. Most products now carry guarantees and there is consumer legislation to support product quality and safety. Therefore, the driving factors in the function of a product are these:

• How easy is the product to use?
• Is the product value for money?

Quality of manufacture

The quality of a finished product is governed by how well it has been made. The development of new materials, manufacturing processes and technologies have greatly increased the quality of many products. Equally so, these new techniques have had a constant influence on the design of products. For example, the process of injection moulding plastics has greatly influenced the design of many different products from the casings of electrical goods to packaging. It is used because it is cheap, quick, efficient and suits a wide range of plastics and plastic products. Injection moulding is suitable for complex shapes with holes and screw threads and for multiple moulds for small products such as small containers and lids.

The mini POLO dispenser is an example of a product that makes full use of injection moulding. The innovative lid mechanism is achieved by the thinning of the plastic to form a sprung hinge which is opened when pressure is applied to a specific area of the casing. The dispenser is formed in two parts with the lid using low-relief lettering for the brand name. The lower half of the casing is filled with mini POLOs and the lid is snapped securely shut. Once this new product was introduced, sales of POLO increased.

> ■ **Things to do** ■
>
> 1 Compile a list of products that you think have been designed well. What is the appeal of these products?
>
> 2 Select a couple of products that you think have been made well. Describe the process of manufacture and how individual parts fit together or work.

Analysing brand identity

Aims

- To understand what makes an effective brand name, logo, symbol or trademark.
- To understand the importance of brand identity in creating the 'right image'.

Modern life is full of **branded goods**, from the cornflakes we eat for breakfast to the trainers we wear in our leisure time. Our choice of brands says a lot about us as individuals. All graphic products have a brand identity and it is important that we analyse this in order for us to design our own.

Logos, symbols and trademarks

Branding goods dates back to the practice of marking property such as sheep or cattle with a hot iron in order to prove ownership. Today, the most recognizable feature of brand is a name, logo, symbol or **trademark** to differentiate one product from another.

Type	Description	
Name only	This developed from the owner of the company putting his or her name or signature to the product to assure quality	*Kellogg's*
Name and symbol	This uses the name of the brand, but incorporates it within a simple symbol such as a circle or square	Levi's
Pictorial name	This is where the overall style of the logo is distinctive and does not need the name to identify it	McDonald's

Types of logo

Typeface	Example	Examples of typefaces (fonts) available in word processing and DTP programs	Description	Image
Serif – typeface has 'tails' at ends of letters	Typeface	Times New Roman, Courier, Bookman	Easy to read, pleasing on the eye	Traditional
Sans serif – typeface has no 'tails'	Typeface	Helvetica, Arial, Eras	Strong, bold and clear	Modern
Script	Typeface	Forte, Mistral, Freestyle, French and Brush Script	Looks personal (handwritten), can be difficult to read	Historical/personal
Decorative	TYPEFACE	Jokerman, Chiller, Victorian, Old English Text	Attracts attention, can be difficult to read	Modern or historical depending upon style

The four main typefaces

Typefaces

The use of the correct **typeface** is very important for any brand name or logo.

Colour

Logos should use a limited range of colours in their design. The most effective logos use a maximum of three colours, but the choice of colour is also very important for the brand.

The British Airways logo uses the red, white and blue colours of the British national flag

In focus – Nike

The best symbols are simple and distinctive and also manage to say something about the nature and quality of the products or services they offer. One of the most famous sports brands is Nike. The company takes its name from the winged Goddess of Victory and the distinctive Swoosh logo is a simplified version of one of her wings. Nike uses several different variations on its main logotype in its products but the Swoosh logo has now become so recognizable that the company does not even have to include the word Nike anymore!

The Nike Swoosh

Analysis – what makes an effective logo?

- What type of logo is it?
- What typeface does it use?
- How many colours does it use?
- What image does the logo create?

Creating the right image

The concept of branded goods is central to our society. When we walk through any shopping area it is covered with images and signs where the brand has become extremely significant to our lifestyle. Many use brands as a means of identification with other people by creating the right 'image'. The product is promoted as a signal to others of our status or personal values. The use of advertising and the media encourages us to buy **branded goods** in order to buy into a particular lifestyle. For example, a person driving a sports car with a BMW badge on the bonnet will automatically be branded successful and wealthy.

Analysing brand image

In order for any brand to be successful it must portray the right image. When analysing the image of any product we should ask ourselves two important questions:

- Who is it aimed at? (The target market)
- What does it say about the people who buy it? (Their lifestyle)

The Nike drawstring bag

The Nike drawstring bag

A common sight in our streets and schools is the Nike drawstring bag. This simple nylon bag with a drawstring closure is used not only to carry sports kit, but books and other personal items. As a result, it is aimed primarily at the youth market. The Nike drawstring bag, however, sends the message that the wearer is aware of fashion and quality goods and is therefore socially acceptable. This is achieved simply by the use of the distinctive Swoosh logo. There are other nylon bags that carry other logos, but these do not send out the same message.

> **▪ Things to do ▪**
>
> 1 Discuss the importance of branded goods in your life.
>
> 2 Cut out or draw examples of logos and analyse them using the information on these pages.
>
> 3 Create a lifestyle analysis mood board for a particular brand.

Analysing marketing issues

Aim

- To understand how to use the 4 Ps – product, place, price and promotion – in order to analyse graphic products.

Marketing is the process of identifying, targeting and satisfying customer needs by the use of marketing objectives, strategies and tactics. It is only by analysing marketing issues that we can design a marketable product.

The marketing mix – the 4 Ps

The marketing mix is a checklist of elements of marketing that companies focus on when carrying out market research and strategy. The marketing mix is often referred to as the 4 Ps – product, place, price and promotion. We can use the 4 Ps to analyse any graphic product.

Product

The product is the most important element but needs the support of all the other factors. Market research will identify a need and product development will create a product in order to best satisfy that need.

Comprehensive market research will identify a target market by:

- knowing who the customer is (age, income, lifestyle, etc.)
- how the customer makes his or her purchasing decisions (advertising, media, product placement, etc.)
- what the customer wants from the product (specification)
- if there are gaps in the market (unsatisfied needs and opportunities)
- analysing what competitors are doing (existing products).

Place

Choosing the right place to sell the product

Where are the best places to sell the product for reaching potential customers?

This may seem quite obvious, for example the best place to sell popcorn is at the cinema. Other places can seem less obvious, for example placing magazines at supermarket checkouts entices shoppers to flick through them while waiting in the queue.

Choosing the best way to get the product to the retail outlet

There are three main ways of distributing a manufactured product to the customer: traditional, modern and direct. The traditional way is best illustrated by your local corner shop which will buy its stock from a wholesaler who, in turn, buys it from the manufacturer. The modern way is used by larger retailers or superstores which may buy directly from the manufacturer. The direct way is being made more popular by the use of online shopping on the Internet. Obviously, the fewer people involved in the chain (channels) of distribution, the lower the price of the product.

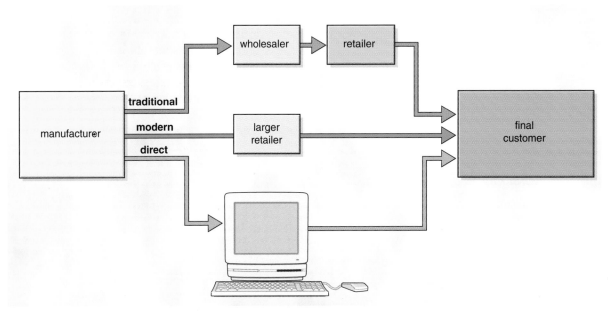

Channels of distribution

Price

Price involves a balance between being competitive and being profitable. It also takes into account what price the target market is willing to pay for the product. If it is too cheap, consumers may think it is downmarket, and if too expensive, they may think it is not good value for money and will not be prepared to buy it.

Promotion

Promotion tells potential buyers about the new product. 'Pull' promotions are aimed at the consumer and 'push' promotions are aimed at retailers, distributors and wholesalers. There is a wide range of promotional strategies including TV, newspaper and magazine (media) advertising, the Internet, billboards and public transport, sponsorship of sports events, point-of-sale displays and direct sales by mailing information direct to your door.

An advertisement on the side of a Beijing bus

Bounty – 'The Taste of the Exotic'

Unique selling proposition (USP)

This approach would take one real feature and make it the most recognized feature in advertising. For example, the Bounty chocolate bar has 'The Taste of the Exotic'. The images that are used to promote it are of tropical beaches, palm trees and, of course, coconuts. By transporting us to exotic locations, Bounty focuses on the fact that it is one of the only coconut chocolate bars on the market

■ Things to do ■

1 Use the 4 Ps to present a written document of the marketing issues involved in a famous branded product, for example Coca-Cola.

2 Discuss the reasons why companies spend millions of pounds on advertising their products.

Design and market influence

111

Analysing packaging

The role of packaging

The role of packaging is extremely important to consider when analysing any package. Packaging has four main functions, as:

- a container
- an advertiser
- a protector
- a preserver.

A container

The package should act as a carrier and a dispenser for the product contained within it. The package has to transport the product safely from the manufacturer, to the retailer and finally to the consumer. The type of package and the materials used in its construction will directly influence the effectiveness of this function. Once in the hands of the consumer, the container has to dispense the product in a safe and convenient way. This can be achieved by a variety of methods including lids, twist caps, aerosols or tear strips.

Analysis

- How does the package contain the product? (Type of package and materials used)
- How does the package dispense the product? (Closures)
- Does the package contain the product effectively? (Any problems with opening and closing the package)

An advertiser

The package should both describe the product and identify it by means of brand identity. The consumer should know what it is he or she is buying in terms of its contents, its weight, and how the product should be used including any special instructions.

Manufacturers will use a brand identity to sell the product. Most products will use printed graphics to make the consumer aware of its presence on the supermarket shelf. Decades of advertising and **brand identity** persuade the consumer to buy the product because they have become familiar and trusted brands.

Faced with the wide choice of similar products, the well-known brand name has consistent quality and the consumer can rely on it. Here, the package can make the choice of product on behalf of the shopper.

Analysis

- Does the package contain all the necessary labelling and easily understood instructions for use? (Food labelling regulations)
- How does the package differentiate itself from its competitors? (Brand identity)

Liquid is easily dispensed by the use of a directional pouring lip and any excess liquid drains back into the container when set upright

Label containing information and communicating brand identity

A protector

The transportation of a product without breakage or spillage is of prime importance and it is the role of packaging to prevent this. The choice of material for a particular product has serious consequences for the effectiveness of the package. For example, glass

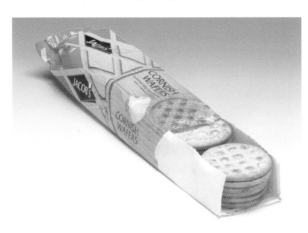

Protecting brittle products

Preserving food with different shelf lives

bottles are an excellent package for fizzy drinks but they are easily broken if accidentally dropped. Some packages such as biscuits have internal packaging in order to protect their brittle contents. This also causes environmental problems with the amount of packaging material used and its disposal after use.

Analysis

- How does the materials used in the package offer protection to the product?
- Could alternative materials be used?
- Could less packaging materials have been used?

A preserver

The package must be able to preserve the product for the necessary amount of time to prevent spoilage. Some products have greater shelf lives than others, for example fresh vegetables may be packaged in a tray covered with plastic film as they have a shelf life of a couple of weeks, whereas, tinned vegetables can keep for months or years due to the cooking and canning process.

Analysis

- How does the package preserve the product? (Processes and materials)

■ Things to do ■

1 Analyse a packaged product using the four main functions of packaging as criteria.

2 Discuss examples of packaging that do not meet these four main functions. What are the major problems in their use and how could these be overcome?

Product disassembly

Aims

- To understand the stages in disassembling a product.
- To understand the component parts, materials, manufacturing and printing processes involved in the production of graphic products.

The complete product and its component parts

You can get a clearer understanding of how a product is manufactured and put together by taking apart an existing product and examining its various parts or components. This will help you to work out:

- the function of its components
- the materials used
- the manufacturing processes used
- the printing processes used.

Disassembling a milkshake bottle

There are a range of flavoured milkshakes available on the market. We are going to look at a bottle of Nestlé's Nesquik chocolate milk. First, we will take the bottle apart in order to determine its component parts.

Stage 1: Working out the function of component parts

Here we can clearly see that the milkshake bottle is made up of four parts: a cap, a seal, a plastic bottle and a plastic label. Now we must work out what each part does.

- The bottle cap provides a means of opening the bottle in order to dispense (pour) the drink.
- The seal provides a means of ensuring that the drink is kept fresh and provides security so that the contents cannot be tampered with.
- The bottle is a container for the drink.
- The label has the brand identity printed on it for easy product recognition and also carries all of the relevant legal labelling.

Stage 2: Working out the materials used

Each component part has been made of a specific material that will ensure it functions as intended. We know that the cap, bottle and label are made of plastic, but which plastic exactly? By looking for the plastics coding symbol on each part, we can check which type of plastic has been used and the reasons for its use by looking at the table on page 17.

Look for the plastics coding symbol on the lid and the base of the bottle

- The cap is made from polypropylene (PP 5) because it needs to be stiff and rigid as it includes a screw thread for opening and closing. It also has good impact resistance so if it were to be dropped, it would not crack.
- The bottle is made from high density polyethylene (HDPE 2) because it is a rigid, tough and hard-wearing material for containing liquids.
- The plastic label has no visible code printed on it, so we must wait until we determine its method of manufacture and then we might have some further clues.
- The seal is clearly made of aluminium foil because it provides an excellent barrier against moisture and can be heat sealed to lock in freshness and prevent contamination.

Stage 3: Which manufacturing processes were used?

By applying our knowledge from earlier in this book regarding the thermoforming of plastics (see pages 16–17), we can work out the method of manufacture for each component part.

- The bottle itself is hollow and has a feint line running down each side and across the bottom. We can work out from this information that the bottle has been formed by a split mould (two halves) and because the bottle is hollow, it has therefore been blow moulded.
- The bottle cap is not hollow but has been formed into a dish shape. The bottle cap also has an intricate low-relief moulded graphic on the top. We can therefore work out that the cap has been injection moulded.
- The plastic label fits tightly around the bottle which is a characteristic of **shrink wrapping**. We would then have to further research shrink wrapping to fully understand the process and the types of plastics used (see below).
- The foil seal has a specific round shape with a tab which was die cut from a large sheet of aluminium foil and heat sealed on to the top of the bottle.

Results of further research

Shrink wrapping using PVC sleeves offers all-round graphic coverage. The bottle is labelled when full. A PVC sheet is rolled and heat sealed to form a tube. The PVC tube is placed over the bottle and an infrared hot air system applies heat (pre-heated to 58°C). The PVC sleeve shrinks between 60 and 90 degrees, moulding itself around the bottle. The bottle is rotated during the process to ensure an even shrinkage all around (shrinkage is around 58 per cent in diameter).

PVC is used for the English market, but some European countries prefer a 'greener' alternative such as PET.

Stage 4: Which printing processes were used?

Both the PVC shrink sleeve and aluminium foil seal have been printed on.

By using the guide to printing on various materials (see page 37), we can work out that both the PVC shrink sleeve and aluminium foil seal were either printed using the gravure or flexographic processes. **Flexography** (see page 36–7) is currently the most widespread of the two printing processes, so it is likely that this process was used.

In conclusion, we can analyse any graphic product by using a combination of reasoning using the information contained within this book, further research and a little bit of guess work.

▪ Things to do ▪

1 Present the information on these pages on one piece of A3 paper. You could use labelled sketches or digital photographs of a similar milkshake bottle to document the disassembly.

2 Disassemble an aluminium drinks can using the four stages described above.

Presenting the results of analysis

Aims

- To understand the value of focused case studies for analysing products.
- To understand the need for clear presentation of results.

Focused case studies

Focused case studies are a useful way of gaining greater awareness and insight into the development and marketing of a particular graphic product.

A good focused case study will include:

- a visual **analysis** or **product disassembly** determining component parts
- a justification of materials used
- a justification of manufacturing and printing processes used including scale of production and quality control and assurance
- marketing issues and product range
- product history and development.

Carrying out research

Obviously, not all the information required will be at your fingertips. It is therefore important that additional research be carried out using a variety of methods including writing or e-mailing companies, visiting websites, visiting companies or exhibitions and speaking to experts.

For example, when conducting a focused case study on Smarties, a student visited the Nestlé website, wrote off to the company for information and e-mailed a marketing director from the company. Most importantly, the student gained first-hand knowledge of the manufacturing processes used in producing the Smarties tube.

Product disassembly

An important aspect of the case study is to carry out a visual analysis of the product by disassembly (see pages 114–15).

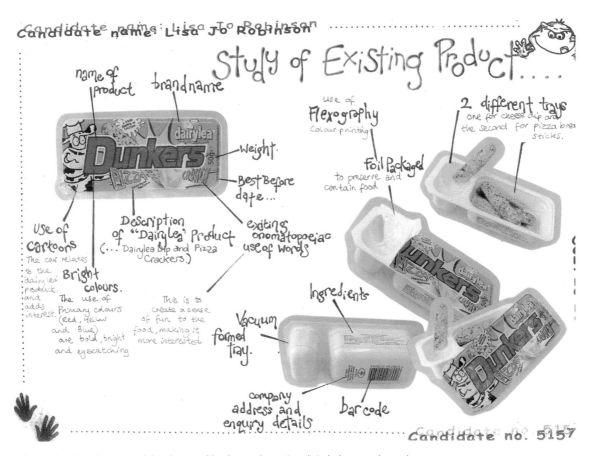

Here the student has documented the disassembly of a product using digital photography and clear labelling of all component parts, materials and manufacturing and printing processes

Design and market influence

116

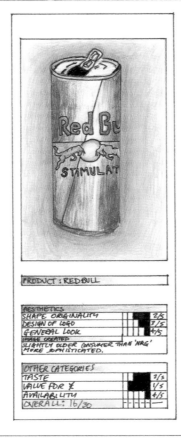

PRODUCT

COMPARISONS

PRODUCT: NRG - LUCOZADE

AESTHETICS
SHAPE ORIGINALITY	4/5
DESIGN OF LOGO	3/5
GENERAL LOOK	4/5
IMAGE CREATED,	

Young, sporty but also appeals to rave scene.

OTHER CATEGORIES
TASTE	3/5
VALUE FOR £	2/5
AVAILABILITY	4/5
OVERALL: 20/30	—

PRODUCT: RED BULL

AESTHETICS
SHAPE ORIGINALITY	2/5
DESIGN OF LOGO	3/5
GENERAL LOOK	4/5
IMAGE CREATED	

Slightly older consumer than 'NRG' more sophisticated.

OTHER CATEGORIES
TASTE	2/5
VALUE FOR £	1/5
AVAILABILITY	4/5
OVERALL: 16/30	—

PRODUCT: DOUBLE 'E'.

AESTHETICS
SHAPE ORIGINALITY	2/5
DESIGN OF LOGO	5/5
GENERAL LOOK	5/5
IMAGE CREATED	

15 - 25, DANCE SCENE, DOUBLE 'E' POSSIBLE REFERENCE TO DRUGS.

OTHER CATEGORIES.
TASTE	3/5
VALUE FOR £	2/5
AVAILABILITY	1/5
OVERALL: 18/30	—

JONATHAN STRASNY

A visual way of comparing similar products

Product comparisons

Product comparisons are a visual way of comparing similar products available on the market. Each product is evaluated using a set of criteria common to all. The results can be communicated as pictograms or short written evaluations.

In the example below, the student has compared three energy drinks. The page is well presented and has been divided into three with a good quality sketch of the product clearly visible. An alternative may be to use a digital photograph. The student has then decided upon a set of common criteria on which to evaluate all three products. The results are presented as a series of charts with a simple rating scheme to communicate the results. In this way, we can clearly see which product the student favours.

Design and market influence

Practice examination questions

1 The specification for a prototype shower gel container is that it must:

- be easy to use while in the shower
- incorporate the brand name 'ZEN';
- be made from inexpensive and easily recyclable materials
- be thermoformed.

This design idea has been proposed by a designer.

a Name the two specification points which have *not* been met by the design and describe how they have not met this point. **(6 marks)**

b Use notes and sketches to show one design idea for the prototype shower gel container which meets *all* the initial specification points. **(8 marks)**

c Explain two factors which must be considered by a designer when designing the shower gel container suitable for mass production. **(4 marks)**

2 A CD-ROM offering free Internet access is to be given away with a computer magazine.

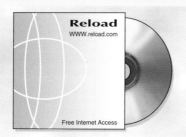

The magazine is an example of a traditional printed medium, whereas the Internet is an example of a new electronic communications technology.

a Name two other electronic communications technologies. **(2 marks)**

b Explain three reasons why the Internet has become so successful. **(6 marks)**

3 Bar codes are used on the majority of printed graphic products.

a Name two pieces of information stored on a bar code. **(2 marks)**

b Describe how the use of a bar code and computer systems enable fast and easy communication of data. **(5 marks)**

4 A company's sales and expenditure graph for a new product is shown below.

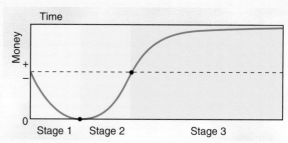

a Study the sales and expenditure graph.

i State the important event that happened at point A. **(1 mark)**

ii Explain the significance of point B. **(2 marks)**

b Describe three things that are happening in Stage 1. **(6 marks)**

c The product has had millions of pounds spent on an advertising campaign.

i Describe three marketing techniques that may be involved during Stage 2. **(6 marks)**

ii Explain how the company may employ the use of an electronic communications technology at this stage. **(3 marks)**

Section E:
Full course coursework

Coursework project design folder

Coursework design folders should be as concise as possible. You should include relevant information only and avoid padding and repetition.

For example, when gathering information from the Internet or a book, select the relevant information only and represent the key points on your page. Avoid simply printing out or photocopying pages of unnecessary information.

The coursework project should be 15–20 A3 pages in total and take up to 40 hours to complete.

Organize your project in a logical manner by numbering all of the pages and include a contents page. This will help your teacher to mark each of the criteria and help Edexcel to moderate your teacher's marks.

Design pages should show a logical progression of design ideas – from the generation of several different initial ideas, through the development of one or two designs, to the final workable design proposal.

Make all major decisions along this design process very clear.

It is extremely important that you include a clear photograph of your finished product. Take several photos and choose the best ones. Photos can also be taken from all angles to give a better impression of the product and to show details.

Above all, be creative and enjoy yourself!

Criteria	Coursework design folder contents	Marks awarded	Page breakdown
1	Identify needs, use information sources to develop detailed specifications and criteria	9	4–5
2	Develop ideas from the specification, check, review and modify as necessary to develop a product	27	6–7
3	Use written and graphical techniques including ICT and computer-aided design (CAD) where appropriate to generate, develop, model and communicate	9	Evidenced throughout
4	Produce and use a detailed working schedule which includes a range of industrial applications as well as the concepts of systems and control. Simulate production and assembly lines using appropriate ICT	9	1–2
5	Select and use tools, equipment and processes effectively and safely to make single products and products in quantity. Use CAM appropriately	39	3–4
6	Devise and apply tests to check the quality of own work at critical control points. Ensure that product is of suitable quality for its intended use. Suggest modifications that would improve the product's performance	9	1–2
	Total	102	15–20

Full course coursework assessment criteria

Criteria 1

Identifying a need

It may be safe to say that as you are undertaking a GCSE in Graphic Products it means that you enjoy designing and drawing. When undertaking your final major project, remember this and adapt your project so that you can design and produce a product that is of some interest to you. We all know that we don't work as well as we might on things that have very little interest to us.

Bearing this in mind, it is extremely important to get the situation correct in the first place. This will set the scene for your design-and-make activity. At this stage, your teacher may actually give you the context so that choosing your own is not so daunting. This is a good thing as having too much choice may mean that you spend weeks trying to think of an area that you want to design in.

For example, the teacher has given the whole class the following situation for their project: *'Brand identity is an extremely important aspect of any product. It makes products instantly recognizable and forms the basis of their marketing and advertising'.*

We must now **analyse** this statement in order to come up with our design brief. For example, Student A plays a lot of sports in and out of school and buys energy drinks in order to replace lost fluids while playing these sports. It might be a good idea for Student A to put the emphasis of his analysis upon energy drinks as this is a personal interest.

Aims

- To recognize a situation for design and to write a detailed design brief that identifies a product and potential users in a target group.
- To select and use data that is relevant to the product and users.
- To analyse research data and develop a specification for testing and evaluating the final product.

How can we analyse?

There are many ways of analysing the situation. These include **brainstorming**, quick surveys or 'straw polls', **mood boards** and attribute analysis charts.

Brainstorming

This can be done alone or with a small group with the same interest. Write the situation statement in the middle of a piece of paper, then note important things to consider before you start designing. These subsections can be broken down even further to help you think about the problem in more detail.

Mood boards

It is essential that you identify a target market for your product early on in the project. This will help you to be selective with your research and aid designing. A collage of images and key statements relating to your target market can set the mood for the project.

■ Hints and tips ■

Your project must have a 2D and 3D outcome. Make sure that any project you use is capable of this.

Lifestyle analysis mood board

Attribute analysis

Attribute analysis is useful for making clear aspects of the context. Start by drawing up a table containing relevant headings about your intended product, for example possible materials, possible size. Then list the possibilities in each column. Highlight the possibilities in each column. This will help you to create different combinations to form statements to aid designing. An example of an attribute analysis chart is shown below.

Possible materials	Possible image	Possible user	Possible scale of production	Possible size	Possible cost
Glass	Sporty	Children	One-off	250 ml	30p–50p
Paper	Sophisticated	Teenagers	Batch	330 ml	50p– £1.00
Board	Cartoon	Early 20s	High volume	1 litre	£1.00+
Metal	Modern	25–35-year-olds		2 litre	
Laminates	Traditional	Mature			
	Hi-tech	Elderly			
	Fantasy				

Attribute analysis for an energy drink project

Writing the design brief

The design brief should be a precise statement of what it is you are going to design and make. It usually starts *'To design and make a ...'*. Some design briefs may be more specific than others depending upon the analysis carried out previously. For example:

- *'To design and make a board game based upon a TV programme'* – this is quite a general or 'open' design brief. We expect a board game to be produced, but we don't know what TV show it is to be based upon, so further research should now focus upon an appropriate show.

- *'To design and produce a concept model and point-of-sale display for a next generation mobile phone aimed at the teenage/early twenties age group'* – this is a more specific design brief. We might expect a 3D model of a mobile phone made out of foam or MDF with a particular styling and function aimed at a specific target market. The 2D aspect is covered by the design of a point-of-sale display for the new phone. Research can now be quite specific looking at the needs of the target market and mobile phones already available to it.

- *'To design and produce the business stationery and a mock-up for the new shop front of the Upper Cut hairdressing salon in Cherry Tree Road, London'* – this is a really specific or 'closed' design brief. The student has identified a real life need and research can now take place with the needs of the client at the forefront. When it comes to reviewing ideas and eventually testing and evaluating the finished product the student will have the advantage of gaining the opinions of a third party.

To be successful you will:

- Identify a realistic need.
- Explore the problem through analysis and investigation.
- Break the problem down into smaller sub-sections.
- Provide opportunity for a 2D and 3D outcome.

Marks awarded: 3

Gathering information for research

Now that you know what you are going to design and make you must gather information from a range of sources in order for you to gain some in-depth knowledge and understanding of your chosen area. This can be carried out in a variety of ways but may include market surveys and analysis of existing products and processes.

Questionnaires

When designing a **questionnaire** it is important to have a clear understanding of what exactly you want to know so that you can ask the appropriate questions. Use closed questions (requiring only yes or no answers) which are easier to analyse and to produce graphs showing the results. For example, how much would you be prepared to pay for a new energy drink?

30–50p ☐   51–80p ☐

81p–£1 ☐   over £1 ☐

This closed question gives definite categories for answers and the results can easily be entered into a spreadsheet package and a graph drawn.

Analysis of existing products

By analysing existing products, you can gain a valuable insight into commercial product design, aspects of which will aid your design. Only select products that are relevant to your project and avoid repeating the process again and again if the information is very similar. **Analysis** of existing products can be achieved in three ways:

- focused case studies
- disassembly of a product(s)
- comparison of similar products.

It is extremely important that you do not spend a great amount of time on this stage of the design process. Research should be relevant and to the point – there is no need for repetition of information.

> ### ■ Hints and tips ■
>
> All information gathered here will directly influence your design specification. It may be useful to include a summary of your research in order for you to think about the writing of your specification.

> **To be successful you will:**
> - Use a wide range of sources.
> - Avoid repetition or useless padding.

Marks awarded: 3

Writing the design specification

This is one of the most important aspects of your project as it will determine the criteria for your finished design. It will form the basis of all your evaluative comments during the design and development stage and the testing and evaluation of the finished product at the end of the project.

The design specification is a list of bullet points that are more specific than the design brief. It must cover all of the important aspects relevant to your design brief including:

- **Function** or user requirements – what will my product have to do?
- **Aesthetics** or appearance – what could my product look like?
- Target market – who will buy my product?
- Materials to be used – what are the best materials to use?
- **Scale of production** and production processes – how many will be made and how?
- Size and cost limitations – does it have to be a certain size and cost a certain amount?
- Legal, moral, social and environmental considerations – does it have to comply with any laws or safety legislation? Could I make it more environmentally friendly?

Look at the example below. Here a dozen specific points make up the criteria for this new product. The design of the product must now take into account all of these points.

> ### Design brief
>
> To design and produce the brand identity for a new unisex perfume called 'Duo' incorporating a prototype bottle, packaging and point-of-sale display.

> ### Specification
>
> The product must:
> - fulfil the general packaging requirements by containing, advertising and protecting the product
> - have a strong brand identity and incorporate the brand name 'Duo'
> - appeal to both male and female customers
> - be aimed at the youth market
> - be made from quality materials which are durable
> - use full-colour printing
> - be able to be mass produced for mass distribution
> - be able to hold 100 ml of perfume
> - incorporate the necessary legal labelling
> - be priced between £10 and £30 to make it competitive
> - address environmental concerns such as use of materials and disposal.

> **To be successful you will:**
> - Write a clear and detailed specification.
> - Include important points of consideration for the proposed design.
> - Provide opportunity for a 2D and 3D outcome.

Marks awarded: 3

Suggested number of pages for Criteria 1: 4–5
Total marks awarded: 9

Criteria 2

Aims

- To present a range of realistic and imaginative design ideas, that relate to the needs identified in the specification.
- To develop, model and test the feasibility of design ideas to produce a realistic design proposal.
- To review design ideas as they develop against the specification criteria.

Generating design ideas

Generating ideas is an extremely important point in the design process. This is the stage where you can demonstrate your individual flare and talent by creating alternative ideas to fulfil the specification.

When generating initial design ideas use graphical techniques that you have developed over your Graphic Products course and which you are comfortable with. For instance, you might find it easier to sketch with a pencil whereas others may prefer an ink pen or even a ball-point pen. The most important thing is to communicate your ideas as best you can. If possible, try to incorporate different materials into your design ideas to show an understanding of more than one material.

There are two main approaches to the presentation of **design ideas** – graphic thinking and more formal drawings – both are perfectly acceptable.

Graphic thinking

Some people find it easy to translate what they are thinking directly into rough sketches to communicate their ideas. Design sheets result in a full and busy appearance with lots of sketches which may overlap each other. The most important designs are highlighted by subtle shading with pencil or markers or fully rendered. The other sketches are as important but fade into the background.

More formal drawings

Other people like to practise drawing their ideas in sketch books and prefer to communicate them as more polished drawings. The appearance of these sheets is more formally structured with the majority of designs in full colour. It may still be possible to include the rough versions of these designs as A4 sheets in your design folder.

When using either technique make sure that all drawings are fully annotated showing any information that cannot easily be seen from the drawings. Additional drawings showing parts of the design in more detail are extremely useful. Most importantly of all, make sure that your design ideas relate directly back to your design specification.

■ Hints and tips ■

A 3D drawing can communicate more information than a flat 2D one. Try to sketch some ideas in 3D by using isometric grid paper for speed.

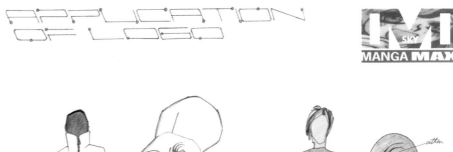

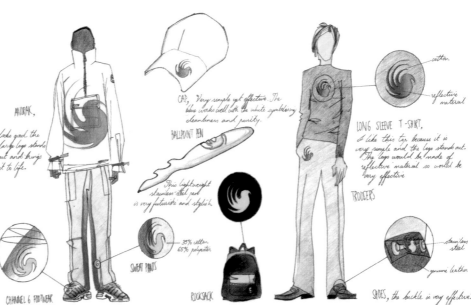

Formal presentation design ideas

Marks awarded: 12

Development of design ideas

From your initial design ideas it will usually be clear which ideas are worth developing and which are not workable and do not meet the specification criteria as successfully.

Developing ideas takes all of the best features of your ideas and attempts to combine them into one workable solution. Specific aspects of the design can be developed individually such as a colour scheme for the final product. Other developmental drawings may show subtle improvements on earlier drafts.

▪ Hints and tips ▪

Further testing and research will help to develop design ideas. In order to make subtle changes to a design, use layout or tracing paper to trace over the previous design.

2D and 3D modelling

At this stage, it may be possible to produce several **mock-ups** of ideas using 2D and 3D modelling. For example, if designing a mobile phone it would be very useful to produce rough polystyrene models to get a feel for the product. It is sometimes difficult to visualize something in 3D just by looking at a 2D sketch. Vital information may be obtained about a design if you can actually pick up a 3D version.

In the example below the student was able to determine the best **ergonomic** design for a radio that could be held easily in the hand and fit into a standard shirt pocket.

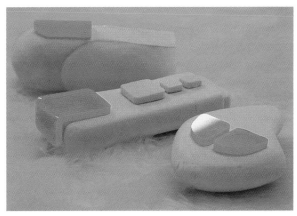

Student's 3D modelling of a hand-held FM radio for promotion of a local radio station

2D or 3D dummies can also be useful to show to a client. These give the impression of what the actual finished product will look like before you go into the specific detail of making. Quick cut-and-paste techniques can be used or the use of computer-generated graphics or images can save a lot of time and can easily be modified.

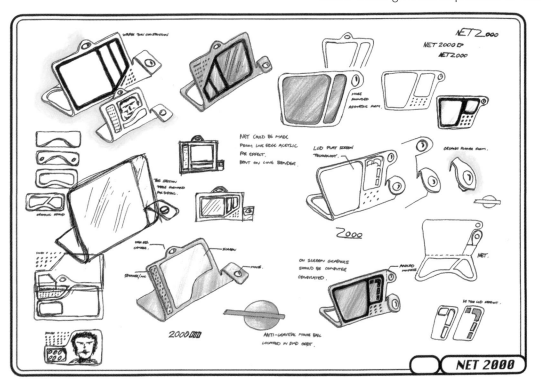

Student's development sheet for an Internet interface showing subtle changes in design

Make sure that you take pictures of all models you have made and include them, along with notes, in your design folder.

Marks awarded: 12

Reviewing design ideas

Reviewing your design ideas can be done in two ways: formatively and summatively.

Formative review

Formative reviewing of ideas should be shown on all design sheets in the form of annotation. Such annotation should show the decisions you have made against the specification.

Summative review

After completing your initial ideas it is a good idea to do a **summative review** of your best ideas. This may be a formally set out piece of work with your best ideas illustrated again, tested against the specification and any advantages and disadvantages of each design explained. At this stage, it is also a good idea to ask other people for their opinions and record the results. Designing should never occur in a 'vacuum' – we always need the views of other people and their suggestions may influence further development of the product.

If you have identified a specific client for whom your product is to be designed, it is essential that his or her opinions are recorded at this stage.

■ Hints and tips ■

In order to test your ideas against the specification you might find it easiest to communicate this information graphically. Use a pictogram for visual impact to rate each design against each specification point.

Marks awarded: 3

Number of pages suggested for Criteria 2: 6–7
Total marks awarded: 27

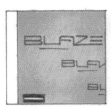

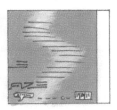

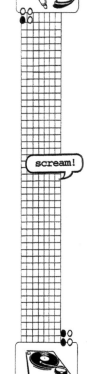

scream!

Spec Point	Rating				
Visual Impact	●	●	○	○	○
Layout	●	●	●	○	○
Information	●	●	●	○	○
Identity	●	●	●	○	○

Not very strong. The lettering is not bold enough and the dull background would not stand out on the shelves.

Spec Point	Rating				
Visual Impact	●	●	●	○	○
Layout	●	●	○	○	○
Information	●	●	○	○	○
Identity	●	○	○	○	○

The hand looks like it wants to stop the blaze. We want the blaze to carry on. The blaze lettering is not very effective over the sign so may not be easily recognisable.

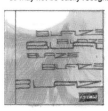

Spec Point	Rating				
Visual Impact	●	●	●	●	○
Layout	●	●	●	●	○
Information	●	●	●	○	○
Identity	●	●	●	○	○

I think that this design is the more effective because of the strong fire emblem. The background could be a bit stronger for effect and the Blaze logo will need to be highlighted so it can be seen. Overall, I think I will develop this one further

Spec Point	Rating				
Visual Impact	●	●	●	○	○
Layout	●	●	●	○	○
Information	●	●	○	○	○
Identity	●	●	●	○	○

I like the flame going white hot. The swirl of the fire and the repetition of the blaze logo could look quite effective.

Summative review of ideas

Criteria 3

Written communication

All Design and Technology focus areas use a technical language that is both precise and unique to that area. There will be a whole list of specialist terminology relevant to Graphic Products and to the design and making of your particular product.

Throughout your project, you must make sure that you use the right name for each material, tool and process and avoid inaccurate descriptions. Write about materials and processes with confidence and avoid using non-descriptive or slang words. For example, 'wooden sticks' are usually referred to as dowels.

All information for your project can be found by using a variety of sources from the Internet to the school or local library, from talking to your teacher to visiting a local company and, of course, this textbook. If you have any doubts about how a particular industrial practice operates, for instance, avoid making your own judgement of how it works and talk to someone who may have the right answer.

The use of written communication will show how much you know about this subject. Read through a description and put it into your own words instead of simply photocopying a section out of a book or downloading masses of text from a **CD-ROM** or the **Internet**.

Marks awarded: 3

Other media

This aspect of the course criteria is important for all Design and Technology areas but more so for Graphic Products as your main material is graphical media. Design projects should demonstrate a range of drawing techniques in both 2D and 3D. Try to include examples of competence in a range of graphic media from pencil rendering to marker rendering, from sketching in pencil to formal drawings using ink pens.

Research and presentation of information

The presentation of research and information should be clear and precise. Sheets should be as full as possible without overcrowding – the use of minimal

Aims

- To clearly communicate ideas and information in a logical and well-organized manner, using appropriate specialist vocabulary.
- To use graphical techniques, models and mock-ups to present ideas and information with a high degree of skill and accuracy.
- To use a range of ICT techniques where available and appropriate.

information and a lot of white space should be avoided.

Large bodies of text can easily be divided up and diagrams and/or photographs inserted. Nobody wants to read essays when a diagram can sum up the information with minimal written explanation. When using photographs, don't forget to add captions to explain fully what the photo is about. Numerical information is best communicated when it is in chart form such as a bar chart or a pie chart.

Designing

The design process requires clear communication of ideas so use a range of drawing techniques as described in Criteria 2.

Working drawings

Working drawings showing vital dimensions and hidden detail (where appropriate) should be drawn in third-angle orthographic. This is a formal drawing technique which should be constructed using a hard 3H or 4H pencil and traced over with black ink pen for best possible effect.

Presentation drawings

A **presentation drawing** should be an accurate representation of your final design solution and should be done to the best of your ability. A variety of techniques and media are useful for this drawing, including isometric, planometric and perspective drawings.

Marks awarded: 3

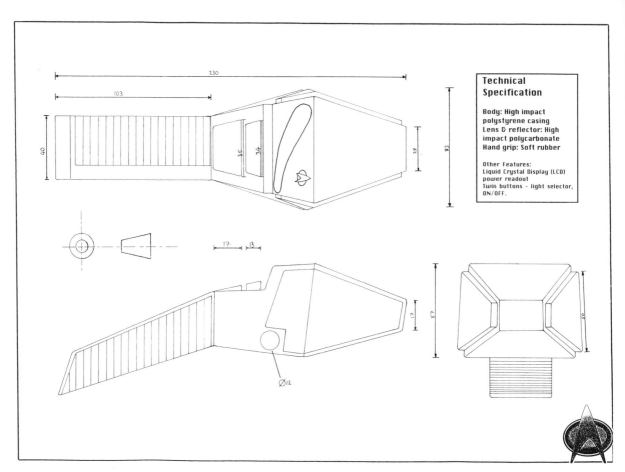

Technical Specification

Body: High impact polystyrene casing
Lens & reflector: High impact polycarbonate
Hand grip: Soft rubber

Other Features:
Liquid Crystal Display (LCD) power readout
Twin buttons – light selector, ON/OFF.

Third-angle orthographic drawing of a student's product

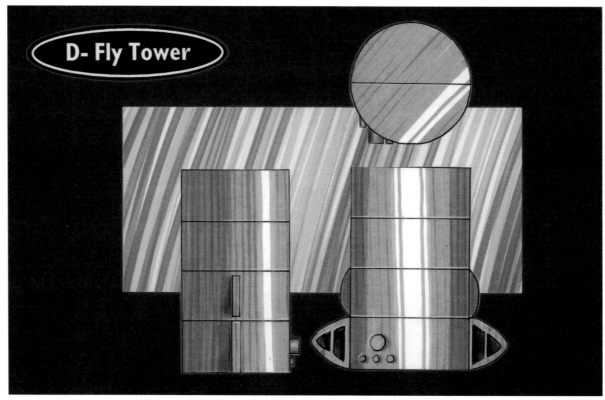

Student's rendered third-angle orthographic used as a presentation drawing

ICT

ICT is an excellent tool for producing high quality graphic products – from simple word processing to professional desktop publishing (DTP) and image manipulation, from 2D draw and paint to sophisticated **3D modelling**. There are a wide range of software packages available to you for creating professional looking printed materials and 3D prototypes.

Researching information

- Use the Internet or a CD-ROM to seek information on materials, technical information and existing products.
- Use **e-mail** to communicate with companies when seeking information.
- Present and analyse information using charts generated in a database or spreadsheet.
- Use a **digital camera** to record disassembly and analysis of existing products.

Generating ideas

- Use 2D draw and paint packages, DTP, CAD and web design software to generate, edit and communicate design ideas.

Developing ideas

- Use a CAD package to select and refine final designs and to produce dimensioned working drawings.
- Use a 3D modelling program to produce a visual image of the proposed completed product.
- Use a spreadsheet to cost a product and to work out the implications of quantity production.

Considering industrial application

- Use a digital camera to record the sequence of making activities and to show processes used in producing your project and/or an industrial visit.
- Use computer-generated **flow charts** to plan the sequence of activities in manufacturing the final product.

Making

- Use a plotter/cutter to produce shapes in card for structural packaging designs and vinyl for stickers.
- Use DTP software to produce printed materials, for example leaflets, menus, business stationery, and so on.
- Output designs to inkjet or laser printers to produce products of repeatable quality.

> **To be successful you will:**
>
> - Demonstrate the use of ICT throughout your project where appropriate.
> - Use a range of software packages in order to communicate ideas effectively.

Marks awarded: 3

Number of pages suggested for Criteria 3: evidenced throughout
Total marks awarded: 9

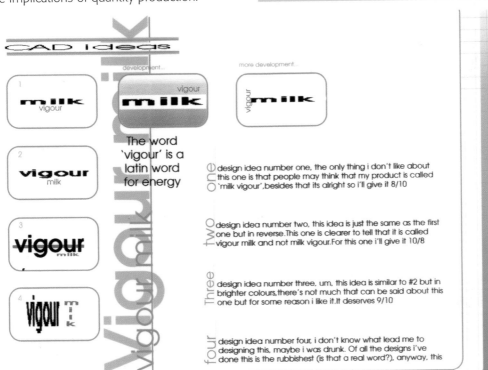

Using draw and paint programs to create logo designs

Criteria 4

Aims

- To produce an outline systems diagram that shows the manufacture of the product. Explain where inputs, processes, outputs and quality control checks occur.
- To produce a working schedule for the manufacture of the product.
- To provide clear evidence of having used appropriate industrial methods of manufacture in your own making.

Systems and control

To ensure careful planning for the making of your one-off product, you must first produce a flow chart showing the various stages of manufacture.

The manufacture of your product involves systems that have **inputs** and **outputs**. Your systems will transform the inputs into outputs. For instance, the inputs in your system will usually be the raw materials, processes and equipment needed to make the components or parts for your product. The outputs will be the manufactured components which will be assembled to produce the finished product.

The **flow chart** should include critical control points (CCPs) where feedback can be identified. This could be stages of manufacture where quality and/or safety checks can be made or specific tolerances are required. A good example of a flow chart will also include all of the various manufacturing sub-systems required to assemble the final product.

■ Hints and tips ■

The planning of manufacture must be completed before you make your product. It is not sufficient simply to include pictures of you making your product as a document of planning. This suggests that you made your product and then sorted out the manufacturing stages afterwards. There should only be sketches of production processes at this stage as your product has not yet been made.

To be successful you will:

- Produce a flow chart including the main stages in the production of a quality product.
- Break down manufacture into sub-systems of inputs, processes and outputs.
- Identify feedback in your system.

Marks awarded: 3

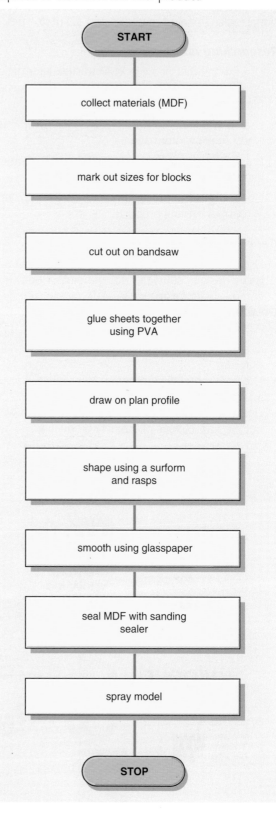

A simple flow chart

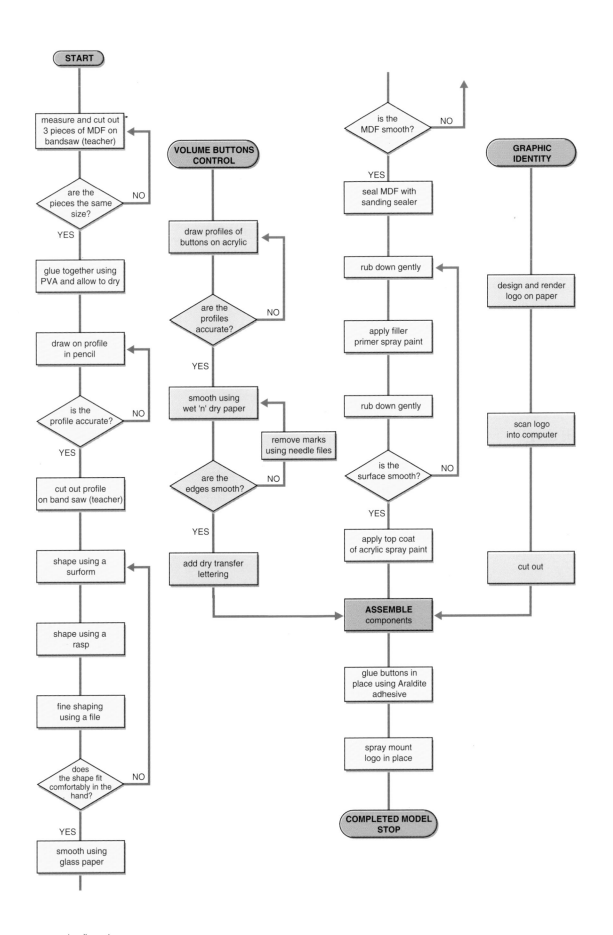

START

measure and cut out
3 pieces of MDF on
bandsaw (teacher)

are the
pieces the same
size? — NO

YES

glue together using
PVA and allow to dry

draw on profile
in pencil

is the
profile accurate? — NO

YES

cut out profile
on band saw (teacher)

shape using a
surform

shape using a
rasp

fine shaping
using a file

does
the shape fit
comfortably in the
hand? — NO

YES

smooth using
glass paper

VOLUME BUTTONS
CONTROL

draw profiles of
buttons on acrylic

are the
profiles
accurate? — NO

YES

smooth using
wet 'n' dry paper

remove marks
using needle files

are the
edges smooth? — NO

YES

add dry transfer
lettering

is the
MDF smooth? — NO

YES

seal MDF with
sanding sealer

rub down gently

apply filler
primer spray paint

rub down gently

is the
surface smooth? — NO

YES

apply top coat
of acrylic spray paint

GRAPHIC
IDENTITY

design and render
logo on paper

scan logo
into computer

cut out

ASSEMBLE
components

glue buttons in
place using Araldite
adhesive

spray mount
logo in place

COMPLETED MODEL
STOP

A more complex flow chart

task		week				
		1	2	3	4	5
1	MDF radio shape					
2	volume/control buttons					
3	graphic identity					
4	assembly					
	time allowed (2.5 hours per week of lessons)	2.5	2.5	2.5	2.5	2.5
					total time	12.5

A Gantt chart specifically for manufacturing stages

Schedule

There are several ways of illustrating a schedule including a **Gantt chart** and a plan to make **storyboard**.

Gantt Chart

A Gantt chart (named after its creator, Henry Laurence Gantt) shows the stages in a project drawn as bars against a time scale. It allows you to view the schedule results set against a calendar. Gantt charts can be used to plan the overall project but more specifically the manufacturing stages.

■ Hints and tips ■

Use a spreadsheet package to create your Gantt charts for speed and effective communication.

Plan to make storyboard

A plan to make storyboard is very similar to a comic strip where each stage of manufacture is drawn in sequence and a caption added. This is a graphical way of illustrating your making intentions and the captions can be very detailed.

To be successful you will:

- Show details of materials and processes identified in your flow chart.
- Consider scale of production, time scale and quality issues.

Marks awarded: 3

Industrial application

Having identified the manufacturing stages for your one-off product, it is now important that you consider its industrial manufacture. During the research stage you should have gathered information about industrial processes. It is now important that you relate the most appropriate processes to your final product as if it were to be produced commercially.

It is important to identify a suitable scale of production for your product as this will considerably alter the processes involved.

Example: high-volume (mass) production

A student has designed and made the **prototype** for a new drinks bottle and has decided that because of the vast amounts of bottles required, high volume is the most suitable scale of production. The bottle is to be made of a suitable plastic, therefore blow moulding is the most appropriate industrial process.

■ Hints and tips ■

Do not simply copy the industrial process out of this textbook but relate it directly to your own product design. For example, if producing a plastic bottle shape it will be manufactured using blow moulding. When illustrating this industrial application, draw your bottle shape and not one you have taken from elsewhere.

To be successful you will:

- Demonstrate an understanding of industrial processes.
- Relate these processes to your final product design and manufacture.
- Consider one-off, batch and mass-production processes.
- Explain changes that may be needed to manufacture in quantity.

Marks awarded: 3

Number of pages suggested for Criteria 4: 1–2
Total number of marks: 9

Criteria 5

Select and use

You will have practised most of these skills during your Key Stage 3 and GCSE Graphic Products course studies. You will have built up a large body of knowledge and understanding about materials, equipment and processes. It is now important that you select and apply the most appropriate techniques in order to produce your own high quality product.

Suggestions on how to provide evidence

Most of this process is carried out by your thinking about the task, trial and error or asking others for advice. This is usually done verbally so you will need to show written or photographic evidence of your decisions in your design folder. A manufacturing report can be a good idea during the making process:

- Keep a weekly diary outlining any problems that you met and how you overcame them.
- Use photographs to show you involved in various stages of production of your product with captions explaining the process and tools being used.

> ### ■ Hints and tips ■
> If things go wrong, don't throw away the results, show evidence of your errors and explain the modifications you had to make in order to put things right.

To be successful you will:
- Demonstrate competence in selecting appropriate tools, equipment and processes.
- Use tools as expertly as possible.
- Show ability to modify techniques and processes where necessary.

Marks awarded: 18

Make product(s)

The most important aspect of this criteria is the production of a high quality product. This can take the form of a prototype or concept model for 3D work and a dummy for 2D work. Remember – the product must include 3D and 2D outcomes.

3D prototypes or concepts

Designing a 3D product will require you to make a realistic model. There are various modelling materials available to perform such a task, but the end result should give the appearance of the actual commercial product.

Aims
- To select and use a range of appropriate tools, equipment and processes with a high degree of skill and accuracy to make a product.
- To apply skills, knowledge and understanding to make a high quality product that fully meets the features of the design proposal. Modify making processes as necessary.
- To use CAM where available and appropriate to improve manufacture.
- To show a high regard for safe working practices.

Example: mobile phone

The student produced this product from scratch by sandwiching together two pieces of MDF, marking out profiles, cutting and shaping. Details and product styling were achieved using an electric multi-purpose tool and buttons were crafted using a soft modelling material which hardens when baked. The screen was computer generated, printed on to transparency and sprayed metallic green on the back. The MDF was coated with sanding sealer so that it would not absorb the two coats of filler primer. In order to obtain a high-gloss effect, several coats of acrylic spray paint were applied. Once assembled, the end result gives a realistic impression of how the actual mobile phone would look.

Student's MDF mobile phone model in open and closed positions

Product Manufacture...

① FIRSTLY, I COLLECTED THICK CARDBOARD TUBING, IN THIS CASE A 'PRINGLES TUBE'. I MEASURED 4 EQUAL PARTS 350MM LONG (DIAMETER 700 MM) I THEN CUT IT WITH A STANLEY KNIFE.

② I SANDED DOWN THE EDGES AND THEN ADDED STRIPS OF CARDBOARD THAT FITTED TIGHTLY INSIDE THE CHUNKERS. FOR EACH PART TO SLOT INTO PLACE

③ I COLLECTED SOME BLUE PVC MATERIAL AND MARKED OUT 3 CIRCLES WITH A SLIGHTLY SMALLER DIAMETER THAN 700 MM, AS THE PVC WOULD BE CUT OUT ON THE FRETT SAW AND THEN GLUED (WITH ARALDITE) INTO PLACE.

(SAFETY GOGGLES USED)

④ AFTER SMOOTHING THE EDGES WITH GLASS PAPER I GLUED THEM INSIDE THE CONTAINERS, AS FLOORS FOR EACH PART.

⑤ I THEN SANDED OFF THE EDGES OF EACH CONTAINER PART, TO ENABLE THEM TO SLOT TOGETHER EASILY AND MAKE IT SAFER TO TOUCH.

Candidate no. 5157

⑥ HERE, I WAS CHECKING TO SEE IF MY MODEL FITTED TO-GETHER NEATLY OF NOT. IF SOME AREAS WERE NOT SO NEAT I SANDED THEM DOWN MORE.

⑦ I DESIGNED MY LABEL ON THE COMPUTER (COREL DRAW) AND ALSO DESIGNED AND PRINTED OFF CAR-TOON CHARACTERS FOR MIX AND MATCH TOY IDEA. WHEN I FIRST PRINTED THESE OFF THEY DID NOT MATCH UP CORRECTLY SO I RE-MEASURED AND RE-SIZED THEM CORRECTLY.

⑧ AFTER GLUEING THE CARTOON STRIPS ON TO THE OUTSIDE OF THE FOUR CONTAINERS, I MADE MOCK UP MODELS OF HOW THE FOOD PRODUCT WOULD FIT INSIDE. (MADE FROM FIMO)

⑨ TO FINISH MY PROTOTYPE I PRINTED MY LABEL ONTO GLOSSY PAPER AND STUCK IT AROUND A SIMILAR SIZED TUBE. THIS IS BECAUSE THE LABEL WOULD ACTUALLY RAP OVER THE CARTOONS ON THE TUBE AND BE TAKEN OFF WHEN BOUGHT. THE LABEL WOULD HAVE A PERFORATED EDGE.

SIDE VIEW

Candidate no. 5157

Documenting the making process using photography and captions

Example: the energy drink

Energy drink prototype can

Alternatively, existing components can be used to produce a realistic model. For achieving the look of a new energy drink product, an actual drinks can was used. The student simply:

- spray painted an empty can
- printed the computer-generated label on to transparency (so the can colour was still visible)
- spray mounted the label on to the can.

2D dummies

With today's printing technology, it is possible to produce dummies that are as close to the actual product as possible. Colour printing from an inkjet or laser printer will achieve good results—especially if a good quality paper is used. Dummies can be made for CD covers, flyers, menus, etc.

CD covers or flyers

For CD covers or flyers, for instance, it is possible to print on to a gloss or satin photo paper and achieve a realistic finish.

Menus

For menus, the paper can be laminated after printing to produce a more durable and wipe-clean surface as used in bars and restaurants.

Use of computer-aided manufacture

Where possible, it is a good idea to demonstrate your competence in using **CAM** to produce your model. From relatively simple vinyl cutters to sophisticated **CNC** routers and milling machines, the use of CAM can demonstrate the production of products of repeatable quality.

■ **Hints and tips** ■

Make sure that you take a clear photo of your product. You may want to include pictures taken from several angles in order to give a better overall impression of your product in 3D.

To be successful you will:

- Produce a high quality 2D and 3D product.
- Produce a fully functional product where possible that meets the requirements of the design proposal and specification.
- Use CAM where possible.

Marks awarded: 18

Work safely

During the making process you should always be thinking about your own personal safety and the safety of others around you.

No one wants accidents to happen and they can be avoided by assessing the risks before you start any making activities. A risk assessment chart will provide evidence that you have taken the necessary steps to avoid any accidents.

During the making process photographic evidence can demonstrate that you have kept to your original risk assessment (this can be combined with your manufacturing report).

To be successful you will:

- Demonstrate a regard for safety awareness for both yourself and others.
- Use risk assessment procedures for the making activities.

Marks awarded: 3

Number of pages suggested for Criteria 5: 3–4
Total marks awarded: 39

Criteria 6

Aims

- To develop and use appropriate tests at critical points to test the quality of design and manufacture against all aspects of the design specification.
- To evaluate the final product in response to the views of intended users and the results of tests and checks made during development and manufacture.
- To use evaluation to suggest and justify modifications to improve product performance.

Tests and checks

Once you have finished making your product, you will be ready to start testing the final outcome against your initial design specification. Some aspects of your product will be easier to test than others (for example where quantitative information can be tested), other aspects will require you to devise practical tests or conduct further surveys of the user or target group.

Write out each one of your specification points and use them as questions. Look at the examples below:

Specification point: The product (packaging) must fulfil the general packaging requirements by containing, advertising and protecting the product.
Question: Did my packaging contain, advertise and protect the product?

By asking yourself this question, you could study your finished product again in detail specifically looking at the function of the package. Tests that could be devised could include making another **mock-up** of the protective cardboard box and determining whether it does protect glass from breakages by using a simple drop test. Photographic evidence of this test should be included in your evaluation.

Specification point: The product must have a strong brand identity and incorporate the brand name 'Duo'.
Question: How did I create a strong **brand identity** incorporating the brand name 'Duo'?

Third-party testing of a student's final product using field testing and a product survey

Candidate name: Lisa Jo Robinson

Questionnaire for Consumers...

	YES...	No...
Must be interesting for young children. Do you think that my 'dairylea' food product design is interesting enough for young children, and would attract their attention well? YES NO	⊦⊦⊦ ⊦⊦⊦ ⊦⊦⊦ ⊦⊦⊦ 20	0
Must fulfil the general packaging requirements by containing, advertising, and protecting the product. Do you think that my design would protect, and preserve the product as well as advertising it well? YES NO	⊦⊦⊦ ⊦⊦⊦ 11 ⊦⊦⊦ 111 18	2
Must have strong brand identity and incorporate the trademark name, 'Dairylea'. Do you think that I have kept 'dairylea's' brand identity constant by utilising a traditional cow cartoon and keeping the background colour blue(on the label)? YES NO	⊦⊦⊦ ⊦⊦⊦ ⊦⊦⊦ ⊦⊦⊦ 20	0
Must be aimed at the ages 5 years+ and appeal to both male and female. Do you think that my package and label design is suitable for this age group and does it appeal to both sexes? YES NO	⊦⊦⊦ ⊦⊦⊦ 1111 ⊦⊦⊦ 1 16	4
Must be safe for young children. Is the package design suitable for young children e.g. no sharp edges? YES NO	⊦⊦⊦ ⊦⊦⊦ ⊦⊦⊦ ⊦⊦⊦ 20	0

The results to this Questionnaire (against specification) for Consumers, are obviously quite positive, which means that overall my product should meet the design specification factors.

MARKETING ISSUES:

Obviously, if this product were to be sold on the market I would display it with the other ranges of 'Dairylea' products in a supermarket.

It would be advertised on television and promoted by the fact that buyers can collect tokens to send off for a promotional t-shirt (which itself would be an advertisement) Also the fact that the container doubles as a toy is a promotion in itself.

Centre no. 14273 3

Candidate no. 5157

A student's questionnaire

■ Hints and tips ■

Allow your target group to view, handle or use the finished product in order for them to gain first-hand experience. This is known as field testing. They will be able to answer your questions more thoroughly as a result.

In order to answer this question, you will first have to determine how to qualify your test results. It is not sufficient to say 'Yes, my packaging had a strong brand identity'. You must show why it had a strong brand identity. This may be achieved by asking other people their opinion or 'third-party' user testing.

A user or target group should have been identified in your specification and a sample from this group could be surveyed. By asking them a series of short questions relating to the effectiveness of the brand identity of the product it should be possible to gain an impression of the market's reaction.

If a client was identified at the start of the project, then this is the ideal point to ask for his or her comments on how well your design meets his or her needs. For example:

'After studying the prototype for the packaging of our own range of salon products, I thought it had a very eye-catching appearance. My staff thought so too. The colour set it apart from other brands so it would attract our customers' attention in the salon.'

Certain specification points will include quantitative information that can be tested easily, for example 'The board game must be no larger than 600 mm x 600 mm'. Obviously, it can be measured in order to answer this point.

Evaluate

Evaluation is an on-going process which should be evident throughout your design folder, from annotation of sketches to formal review of ideas. The final evaluation should consider all these comments and offer an honest opinion of the final product. Quality issues are very important at this stage and will need to be discussed.

To be successful you will:

- Test the finished product against each point of the specification.
- Use third-party user testing where possible.

Marks awarded: 3

Full course coursework

137

No design is perfect, so do not presume that yours is! Be honest with yourself and describe what you do not like about your design. This is not a negative exercise, but it is intended to identify the weak points that would need to be addressed if the design were to be taken any further.

In addition, you will have to include a description of any problems that you met while making the final product. Again, you would have been incredibly lucky not to have had anything go wrong so now you must describe how you overcame these problems. This is an extremely positive aspect of your evaluation as it shows your flexibility in responding to problems by altering manufacturing processes including the use of different materials, tools and equipment.

On the other hand, there will be very positive aspects of your designing and making skills that will need to be highlighted. Things that went particularly well and that you are proud of should be explained in detail to illustrate your knowledge, skills and understanding.

final design idea or your making processes if you had the luxury of starting the project all over again. Suggestions for improvement should relate to:

- product performance
- quality of manufacture and design
- fitness-for-purpose
- target market
- larger-scale production.

For example, in your evaluation you might have identified that your design involved the production of several identical components which you had to make by hand but when assembled they did not fit together perfectly. From this inaccuracy, you may state that the production of several identical components would be better done using **CAD/CAM**. Therefore, in an ideal world and if you had the time all over again, you would draw the component using 2D design software and output this information to a **CNC** milling machine. The end result would also make the product ideal for batch or mass production.

To be successful you will:

- Evaluate work objectively against the specification as designs develop.
- Include problems met and how these were overcome.
- Use third-party evaluation in addition to own assessment where possible.

Marks awarded: 3

Modifications

By testing your finished product against the specification, evaluating the good and bad points and by surveying your target group, you will have identified areas for further improvement.

This last section of your evaluation will suggest any modifications that you would make to either your

---

■ Hints and tips ■

It may be advisable to word process the evaluation so that you can revise and edit it if necessary. Using a computer may also enhance the presentation of this document and help the reader to understand it fully. Use diagrams instead of long explanations where appropriate – 'a picture paints a thousand words'.

To be successful you will:

- Use feedback from testing and evaluation to suggest improvements to both design and making.

Marks awarded: 3

Number of pages suggested for Criteria 6: 1–2
Total marks awarded: 9

Section F:
Short course coursework

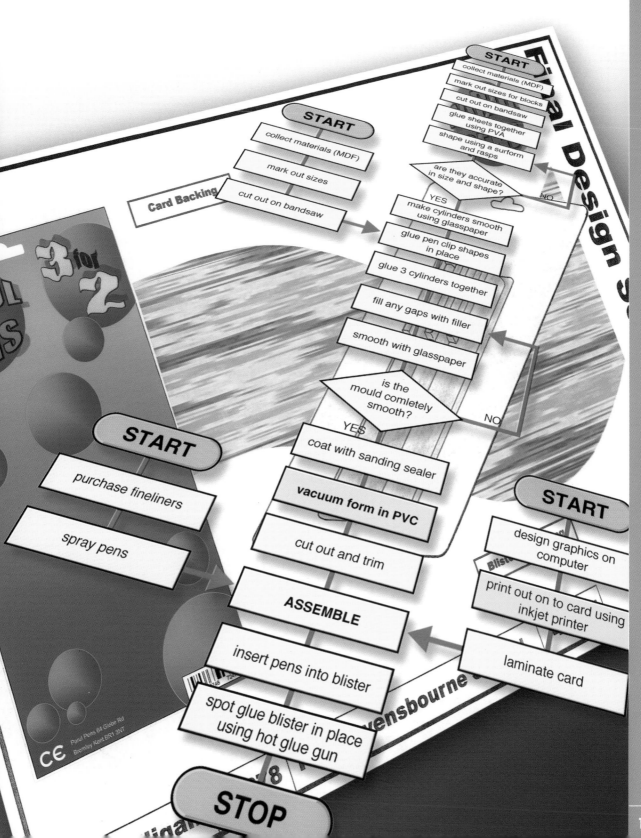

Coursework project design folder

You are required to submit a single coursework design-and-make project that consists of a design folder of approximately ten A3 pages and a practical outcome.

You will be guided to choose one Edexcel-set task from a list of 15 provided each year. Edexcel will also provide you with the A3 proforma pages on which to set out your project. It is expected that you will spend up to 20 hours in producing this project including the making of the practical outcome.

It is essential that coursework design folders be as concise as possible. You should include relevant information only and avoid padding and repetition. The ten-page proforma should be treated as a design exercise in itself in order to include as much relevant information as possible in the space available.

You must be selective when deciding on the content of design folders. You will be allowed to include one supplementary sheet of folio work if necessary.

The front cover of your project must have your Candidate Mark Record Sheet (CMRS) attached. The CMRS should be stapled (not glued) to your project so that the Edexcel moderator can remove it during the moderation process. It is extremely important that you also include clear photographs of your finished product. These must be glued to the front cover.

Above all, be creative and enjoy yourself!

Criteria	Coursework design folder contents	Marks awarded	Page breakdown
1	Use information sources to develop detailed specifications and criteria for a given task	6	1
2	Develop ideas from the specification, check, review and modify as necessary to develop a product	24	4
3	Use written and graphical techniques including ICT and computer-aided design (CAD) (where appropriate) to generate, develop, model and communicate	6	Evidenced throughout
4	Produce and use detailed working schedules, which include a range of industrial applications as well as the concepts of systems and control Simulate production and assembly lines using appropriate ICT	6	1
5	Select and use tools, equipment and processes effectively and safely to make single products and products in quantity. Use CAM appropriately	36	1
6	Devise and apply tests to check the quality of your work at critical control Ensure that own product is of suitable quality for its intended use. Suggest modifications that would improve the product's performance	6	2
	Total	84	9 pages + cover sheet = 10 pages

Short course coursework assessment criteria

Criteria 1

Example of Edexcel-set task

A major stationery company called PARUL PENS is to market blister packs of three ball-point pens at special three-for-two prices. The packs will contain one of each of blue, black and red pens.

Your task

Research the design of blister packs: materials, manufacture, package information, lettering styles, colour, display methods, etc.

Design and make

A high quality blister pack for three ball-point pens on a three-for-two promotion including:

- a full-colour printed card backing
- a vacuum-formed blister shape.

The final design for the blister pack must be suitable for high-volume production.

Aims

- To select and use research information that is relevant to the product and its potential users.
- To analyse research and develop a specification for testing and evaluating the final product.

The Edexcel-set task will always include the opportunity for both a 2D and 3D outcome. Here the 2D aspect is the **design** and production of the printed backing card and the 3D aspect is the design and production of a **mould** for **vacuum forming** the blister shape. Additional work may include the design and making of a further 3D outcome in the form of the pens themselves.

Information gathering

You should consider some of the following sources when collecting information:

- **analysis** of existing products
- consumer surveys – **questionnaire** of target market identifying needs
- visits to manufacturers or using information from their websites.

The information section should be a little under a page of A3 in length so it must be extremely focused. It is a summary of the essential research you have gathered for use when designing your product.

In our example, the most appropriate way of gathering information about blister packs was for the student to disassemble an existing blister pack in order to see how it was made.

Disassembly and analysis

Here we can see that the student has purchased two sets of pens contained in blister packs and has used photographic evidence to document their disassembly. The student has identified most of the processes involved in the manufacture of the product by using annotation and diagrams.

The backing card was made of a medium weight card and the lithographic printing process was used to print the full-colour image. The student then stated why lithography was the most appropriate printing process. The backing card also had a die-cut hanging hook for display purposes. The graphics were bright and colourful and would appeal to a youth market. Important information on the backing card was labelled for easy reference.

The blister shape was identified as being made from PVC because it was transparent and could be thermoformed. The manufacturing process used to form the shape was vacuum forming and a simple diagram was added to illustrate this process.

When removing the blister shape from the card backing, the student found that it was easily removed because of the use of an adhesive similar to hot melt glue instead of a more permanent adhesive. The shape of the blister form itself directly reflected the shape of the pens so that they didn't move about in the packaging.

Sources of information

On this sheet you must list all the sources used to gather useful information on your chosen task. Using the **product disassembly** as a basis for investigation, the student was able to research each stage of manufacture by using textbooks to look up materials and processes. A simple list of all sources used was added to the page as a bibliography.

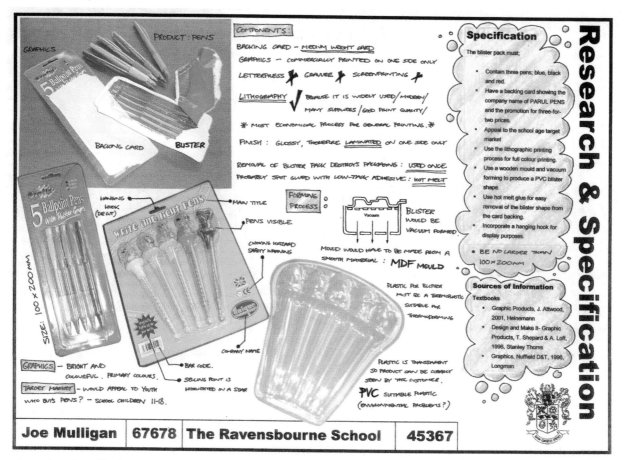

Research and specification

Specification

As a result of the student's investigation into blister packs, a **specification** was drawn up. The blister pack must:

- contain three pens – blue, black and red
- have a backing card showing the company name of Parul Pens and the promotion for three-for-two prices
- appeal to the school-age target market
- use the lithographic printing process for full-colour printing
- use a wooden mould and vacuum forming to produce a PVC blister shape
- use hot melt glue for easy removal of the blister shape from the card backing
- incorporate a hanging hook for display purposes
- be no larger than 100 mm x 200 mm.

To be successful you will:

- Include relevant information only.
- Include a list of all sources used.
- Write a detailed design specification to help you both design a high quality product and write a detailed evaluation.
- Include the opportunity for a 2D and 3D outcome.

Number of pages: 1

Gathering and presenting research information: 3 marks
Developing the specification: 3 marks

Criteria 2

Aims

- Present a range of realistic and imaginative design ideas that relate to the initial design specification.
- Develop, model and test design ideas resulting in a realistic design proposal.

Ideas

This is a very important point in the **design** process and gives you the opportunity to use your creativity to present alternative solutions to the problem and to display your design ability.

At least three different **design ideas** should be presented in this part of the design folder and each design should consider and meet the points you have made in your **specification**. Each idea should be realistic and workable and should be evaluated formatively (or given a **formative review**) to assess its potential.

In our example, the student has presented four different ideas for the backing card. Ideas are all based around the dimensions of 100 mm x 200 mm as directed in the specification and alternative graphics are explored.

The major aspects of the backing card have been identified as:

- company name Parul Pens
- three-for-two sales promotion
- essential information such as the bar code, manufacturer's address and European Safety Standards mark

- background graphics.
- Various backgrounds are evident and each is evaluated for visual appeal. Potential problems are identified and possible solutions offered.

This is a good example of a busy sheet showing that the student is involved in a graphic thinking exercise. The freehand pencil sketches use colour in order to highlight important areas of the page and to reinforce annotation concerning visual impact.

Develop, model and test design ideas

Development brings together the best features of your initial ideas into a final solution that best fits the specification.

At this stage, you may have to compromise some of your ideas in order to allow for cost constraints or the limited availability of materials, equipment or processes available to you in your school.

Modelling and testing your ideas at this point will enable you to see whether your design will work or not. Modelling techniques can be used to produce **mock-up** designs and may include the use of materials and processes that are easier to use than those that will be used in the actual production of the final product. It is important that you record all of your models as evidence of your work.

In our example, the student has used two sheets (including supplementary sheet) to develop both the card backing and the mould for the blister pack. Developmental work for the **mould** is shown using isometric freehand sketching to show the idea in 3D. Useful dimensions are explored and used to develop the mould still further. The backing card is explored through computer-generated graphics focusing upon colour schemes.

Reviewing ideas

Annotation of design ideas should relate back directly to your specification. The student has used a formal review of ideas for the choosing of the colour scheme. By asking three relevant questions the student has illustrated that one colour scheme is more effective than the other two, therefore this one will be developed into the final design solution.

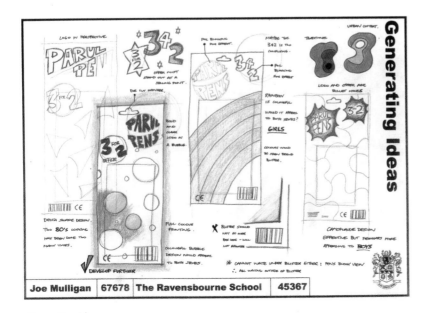

Generating ideas

Final design proposal

Your final developed idea should be presented with detailed information on all points given in the specification with reasons for your decisions.

In our example, the student has presented the final design solution in two parts. The backing card is computer generated and shown actual size with all of its realistic details added. The blister pack is illustrated using a formal isometric drawing showing the position of the PVC blister on the backing card. The two illustrations are connected using a background graphic device to give a well-balanced sheet.

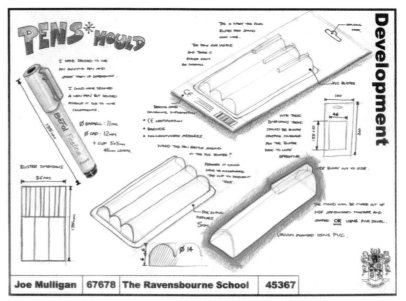
Development of ideas

■ **Hints and tips** ■

This is a very important section in your project and you would be well advised to use the supplementary sheet in order to fully explore your design ideas.

To be successful you will:

- Annotate all design pages.
- Review your designs against your original specification.
- Show a logical progression from initial ideas, through development, to the final workable design solution.

Number of pages: 3
(+ supplementary sheet = 4)

Generating design ideas: 12 marks
Developing, modelling andtesting ideas: 12 marks

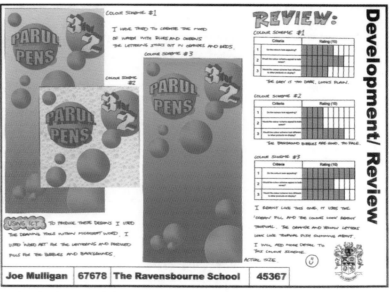
Further development of ideas

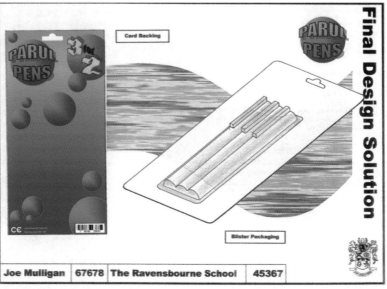
Final design solution

Criteria 3

Aims

- To clearly communicate ideas and information in a well-organized and logical manner, using specialist vocabulary.
- To use graphical techniques, ICT and models to present information and ideas.

Written communication

The study of Graphic Products uses a technical language that is both precise and unique. You should try to use accurate terms when writing up your coursework project. Ideas and information need to be laid out in a well-organized and logical manner.

In order to achieve high marks your work will need to use specialist terminology. It should be used in a well thought-out and logical way.

Other media

In order to fulfil Criteria 3 you must also use a range of media and graphical techniques throughout your project.

You will have to demonstrate good quality communication skills that will show your design ideas in detail. These may include:

- effective and clear sketching techniques
- formal technical drawing, for example third-angle orthographic
- pictorial drawing techniques, for example **isometric**, one- and two-point perspective
- exploded views to show assembly and constructions
- marker and/or pencil rendering
- model making, for example sheet modelling using paper or card, block modelling using Styrofoam or MDF.

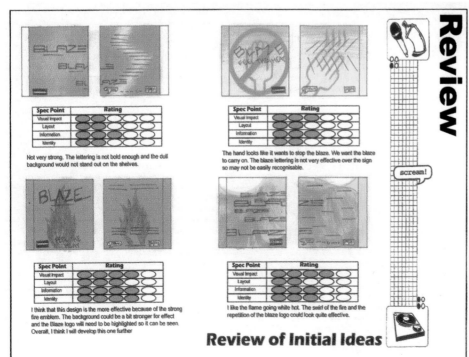

Development showing good use of ICT

ICT

ICT should be seen as another communication tool to be used, where appropriate, throughout your project. ICT can be used to enhance the content and presentation of your folder work. The use of word-processing packages for written work is advisable especially if your handwriting is not very legible. Desktop publishing and spreadsheet packages should also be widely used to present information.

If you use a computer to aid your design work, then you must include details of the software used and a brief explanation of the process used to generate the image.

The following are appropriate uses of ICT in your project.

Researching information

- Use the **Internet** or a **CD-ROM** to seek information on existing products, materials and processes.
- Use **e-mail** to communicate with experts.
- Present and **analyse** information using charts generated in a spreadsheet.
- Use a **digital camera** to record the disassembly and analysis of existing products.

Generating ideas

- Use 2D draw and paint packages, DTP and **CAD** to generate, edit and communicate design ideas.

Developing ideas

- Use a CAD package to select and refine final designs and to produce dimensioned working drawings.
- Use a **3D modelling** program to produce a visual image of the proposed solution.
- Use a spreadsheet to cost a product and to work out the implications of manufacturing in quantity.

Considering industrial application

- Use computer-generated **flowcharts** to plan the manufacturing stages before you begin making your product.
- Use a digital camera to record the sequence of making your product and to document industrial visits.

Making

- Use a cutter/plotter to produce shapes in thin materials for structural packaging designs and vinyl stickers.
- Use DTP software to produce printed materials, for example leaflets, menus, business stationery.
- Output designs to colour inkjet or laser printers to produce products of repeatable quality.

> ### To be successful you will:
> - Show evidence of a variety of drawing styles.
> - Use ICT where appropriate.
> - Use accurate specialist vocabulary.

Number of pages: evidenced throughout

Quality of written communication: 3 marks
Use of graphical techniques and ICT: 3 marks

Criteria 4

Aims

- To produce outline systems diagrams that show where inputs, processes, outputs and quality control occur.
- To produce a working schedule for the manufacture of a product.

Systems and control

Careful planning for manufacture is essential and you should be able to produce a detailed working schedule that would enable another person to make your final product. The basis of a schedule could be a **flow chart** that includes the main stages of production including:

- collecting components and materials
- the preparation of materials when measuring and marking out
- processing and finishing.

You should be able to identify in your work any processes that will be used to shape and transform the **inputs**. Feedback in your system is important and should identify any potential problems or **quality control** points.

In our example, the student has produced a fairly detailed flow chart that documents the main system for the making of the blister **mould** and identifies several sub-systems that feed into the main system at certain stages.

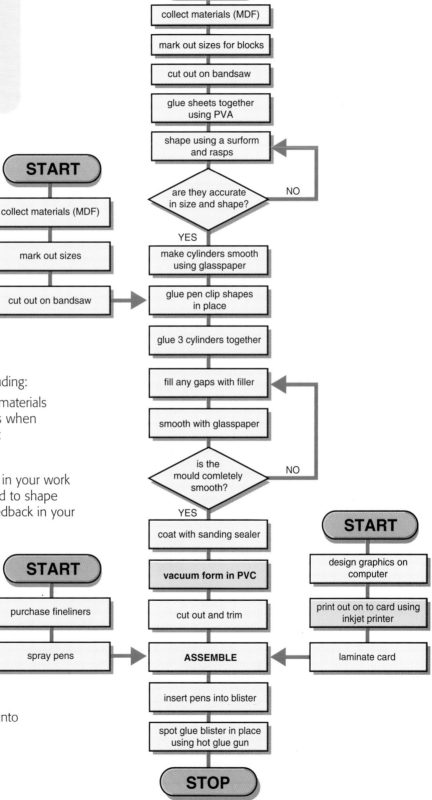

Systems and control flow chart

In order to demonstrate time management skills at this stage, include a GANTT chart detailing the stages of manufacture against the time allocated for making.

Industrial applications

You should demonstrate an understanding of industrial processes and use them in your work where appropriate. Having planned the production of a one-off product, you might consider the demands on equipment and processes of using **batch production** to produce a few hundred of the same product.

In our example, the student has identified two main stages in the manufacture of the **prototype** as ideal for batch production processes. The first is the vacuum forming of the blister pack. The mould could easily be used for producing shapes of repeatable quality. The industrial **vacuum forming** process is essentially the same as that of the vacuum formers commonly found in school workshops. The obvious difference is that industry would use multiple vacuum formers, therefore requiring multiple moulds. The

moulds would therefore have to be more durable and the development costs would be much greater.

The second industrial application is the printing of the backing card. Offset litho was correctly identified as the most suitable process for this and a simple explanation of the process is offered. An example of a commercially printed and laminated piece of advertising is also included to give an impression of the quality of the final product.

To be successful you will:

- Draw up a flow chart for the manufacture of your one-off prototype product.
- Describe the industrial manufacturing processes involved in producing batches of your product.

Number of pages: 1

Producing a detailed work schedule and flow chart: 3 marks
Consideration of industrial processes: 3 marks

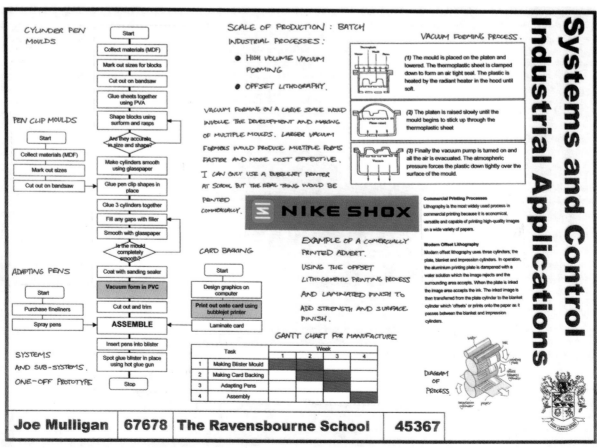

Planning for manufacture: one-off prototype and batch production

Criteria 5

Aims

- To select and use a range of appropriate tools, equipment and processes with a high degree of skill and accuracy to make a product using safe working practices.
- To use CAD/CAM where appropriate.
- To apply skills, knowledge and understanding to make a high quality product.

Select and use

You should show that you are able to select appropriate tools, equipment and processes with reference to your planning for manufacture schedule.

You should use tools and equipment as expertly as possible and should show that you are able to modify techniques and processes when necessary by using your understanding of the limitations and flexibility of tools and equipment.

In our example, the student has attempted to make the mould as planned but has had trouble in creating an accurate shape. Another method of manufacture was explored using different materials and processes based upon the student's knowledge of the subject.

■ Hints and tips ■

Don't give up if things appear to go wrong! There is always an alternative way of making something so ask for advice from your teacher.

Make product(s)

You should aim to produce a high quality product that is complete and fully functional, and meets the requirements of your specification.

The making process gives you the opportunity to demonstrate your skills in using tools and equipment and in applying processes creatively and safely. High

Stages of manufacture storyboard

The student's completed blister pack and pens

quality outcomes demand high level skills not only using processes familiar to you but exploring unfamiliar techniques if appropriate. Challenging work of good quality may be rewarded more highly than high quality work that was more straightforward to produce.

If your school has **CAM** facilities, then you should aim to use them if it is relevant to your project.

In our example, the student has expertly used digital photography to document the main stages in the manufacture of the mould. All problems associated with the original plan of manufacture have been overcome and an alternative means of making was chosen.

All the main manufacturing stages have been photographed with the appropriate comments added. This shows a clear and logical explanation of the making process. Other comments have included safety awareness such as **COSHH** regulations when using solvent-based adhesives and general machine precautions for use of the power saw and vacuum forming machine.

Safety awareness

You must show that you are aware of the importance of safety both for yourself and others when working with materials, tools and equipment. Safety awareness could be recorded in the planning for manufacture schedule or in the manufacture **storyboard**.

To be successful you will:

- Show evidence of making using photographs.
- Produce a product to the best of your ability.
- Include reference to safety awareness.

Number of pages: 1

Ability to select and use tools, equipment and processes effectively: 18 marks
Quality of final product: 18 marks

Criteria 6

Aims

- To use appropriate tests to check the quality of design and manufacture against all aspects of the specification.
- To evaluate the final product in response to test results and the views of intended users and suggest modifications to the product.

Tests and checks

Testing a product against its specification measures its fitness-for-purpose and gives guidance for further improvements and modifications. There are three ways of testing a product:

- testing during manufacture
- testing against the original specification
- third-party user testing – **questionnaires**, field tests, etc.

In our example, the student has addressed all the points in the original **specification** by listing each point and devising tests to find out whether the final product has satisfied them. For example:

Specification point: The blister pack must have a backing card showing the company name of PARUL PENS and the promotion for three-for-two.

This is a relatively easy point to test as it can be addressed visually. A full explanation is given instead of a simple 'yes' or 'no' answer:

'The backing card shows the company name of PARUL PENS in large orange and red letters that can be easily seen against the blue background. The promotion is also in large letters so that customers cannot miss it. The orange and yellow lettering can be easily seen against the blue background.'

Specification point: The blister pack must appeal to the school-age target market.

This point has been tested by the use of a simple survey to the appropriate target market:

'Personally, I think that the design is appealing to the target market because it is bright and colourful and would catch their eye in a stationery shop. To test this theory fully I have asked 20 of my school friends their opinion. I chose 10 boys and 10 girls from years 10 and 11 undertaking their GCSE studies so they regularly buy pens for school.'

A few simple questions were asked in order to find out the appeal of the final product to the target market and the results were illustrated using pie charts.

Specification point: The blister pack must use hot melt glue for easy removal of the blister shape from the card backing.

In order to test this point, the student devised a practical test and recorded the results by using digital photographs:

'I did use a hot melt glue gun to fix the blister to the card backing. I tested how easy it was to remove by vacuum forming another blister and glue gunning it to a similar piece of card. The result as shown in the photo shows that it was quite easy to remove. The card backing would be destroyed and thrown away.'

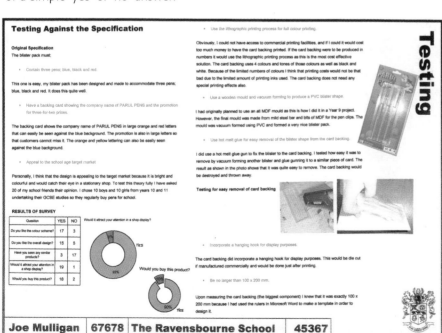

Testing the product against the specification

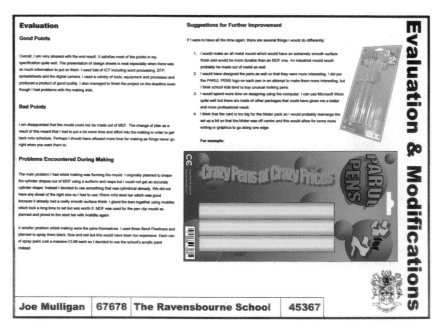

Evaluating and suggesting modifications

Evaluate

You should objectively evaluate your finished product against your original specification in order to justify and confirm its success.

In our example, the student has started the evaluation with a list of 'good points' and 'bad points'. This enabled him to discuss the positive and negative aspects of the project – remembering that no design is perfect.

Your evaluation should include a mention of problems you met, how they were overcome and any modifications made during manufacture.

In our example, the student discussed his problems when making the product and was able to justify the changes he made to the original plan of manufacture:

'The main problem I had whilst making was forming the mould. I originally planned to shape the cylinder shapes out of MDF using a surform and rasps but I could not get an accurate cylinder shape. Instead I decided to use something that was cylindrical already. We did not have any dowel of the right size so I had to use 15 mm mild steel bar which was good because it already had a really smooth surface finish. I glued the bars together using Araldite which took a long time to set but was worth it. MDF was used for the pen clip mould as planned and joined to the steel bar with Araldite again.'

Third-party evaluation would have been a valuable addition to this personal assessment of the student's work.

Testing and evaluating should provide feedback on the effectiveness of your final solution and allows you to suggest design improvements and modifications for future development of the product.

Suggestions for improvement should relate to:

- product performance
- quality of manufacture and design
- fitness for purpose
- target market
- larger-scale production.

In our example, the student has made a series of points that would be addressed 'If I were to have all the time again'. This included:

'I think that the card is too big for the blister shape so I would probably rearrange the set-up a bit so that the blister was off-centre and this would allow for some more writing or graphics to go along one edge.'

This point is reinforced by including a modified card backing in the evaluation.

> ### To be successful you will:
> - Test the final product against each point of your specification.
> - Describe the good and bad points of your final product.
> - Suggest improvements if you were to start the project over again.

Number of pages: 2

Testing the final product against the design specification: 3 marks
Using results of tests to evaluate the product and suggest further improvements: 3 marks

Hints and tips for coursework (short course)

Cover sheet (1 page)

The front cover of your completed project must include:

- the Candidate Mark Record Sheet (CMRS) stapled to the cover so that it can be removed by Edexcel's moderators
- a clear photograph(s) of your practical outcome
- the outline of the short course set task.

Criteria 1 (1 page)

Most importantly:

- Include relevant information only.
- Include a list of all sources used.
- Write a detailed design specification to help you both design a high quality product and write a detailed evaluation.

Criteria 2 (3 pages plus supplementary sheet, making 4 pages)

Most importantly:

- Annotate all design pages.
- Review your designs against your original specification.
- Show a logical progression from initial ideas, through development, to the final workable design solution.

Criteria 3 (evidenced throughout)

Most importantly:

- Show evidence of a variety of drawing styles.
- Use ICT where appropriate.
- Use accurate specialist vocabulary.

Criteria 4 (1 page)

Most importantly:

- Draw up a flow chart for the manufacture of your one-off prototype product.
- Describe the industrial manufacturing processes involved in producing batches of your product.

Criteria 5 (1 page)

Most importantly:

- Show evidence of making using photographs.
- Produce a product to the best of your ability.
- Include reference to safety awareness.

Criteria 6 (2 pages)

Most importantly:

- Test the final product against each point of your specification.
- Describe the good and bad points of your final product.
- Suggest improvements if you were to start the project over again.

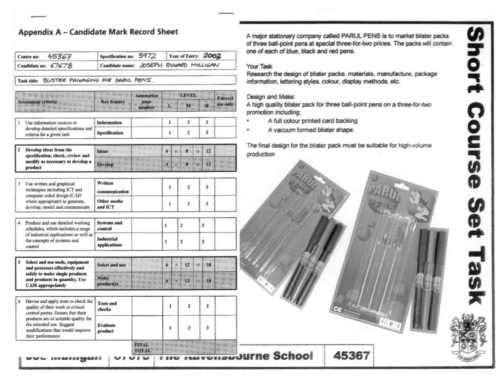

Project cover sheet

Glossary

aesthetics how we respond to the visual appearance of a product, in relation to its form, texture, smell and colour

alumina the material which is formed as a result of refining bauxite (aluminium ore)

analyse/analysis to ask questions using 5WH – who, what, why, when, which (and how)

anthropometrics the study of the human form in relation to size, movements and strengths. Used in ergonomics

antique a term used to describe rough paper used in book printing

attribute analysis identifying the key characteristics, design features of a product

automated a process which is controlled by machine and/or computer

batch production a method of production where a number of components are made all at once. Repeated batches are sometimes made over a longer period of time

blast furnace a vessel in which iron ore, lime and coke are heated to create pig iron (almost pure iron)

bleach a substance which removes colour from a material – used in paper making

blow moulding a process where a thin tube of plastic (parison), gripped between two halves of a mould, is blown out to fill the mould using compressed air. Used for making bottles

brainstorming a group exercise for generating a wide range of ideas related to problem solving. The rule being that all ideas, no matter how good or bad is jotted down on a piece of paper for evaluation

brand identity where a product is instantly recognized as belonging to a particular manufacturer by its image

branded goods a product with a recognized image of quality and desirability

BSI British Standards Institute

CAD Computer Aided Design

CAID Computer Aided Industrial Design

calendering a process where steel rollers are used to improve the surface finish of paper

CAM Computer Aided Manufacture

CAMA Computer Aided Market Analysis

CD-ROM Compact Disc – Read Only Memory

CEN Comité Européen de Normalisation

chemical pulping a process used to separate wood fibre from lignin. Used in paper making

CIM Computer Integrated Manufacture

CMM Coordinate Measurement Machines

CMYK Cyan, Magenta, Yellow and Black. Process colours used in printing

CNC Computer Numerical Control

coated a surface finish applied to a material, like a varnish

coloured dyes chemicals which added to a material alter the colour

composites materials made up of more than one base material, some times in layers or as a mixture

computer models digital images of a design seen on a computer, sometimes as fully rendered products or wire frame images

conversion of timber the process of preparing wood for use as a construction material by seasoning and cutting

COSHH Control of Substances Hazardous to Health

dandy roll a type of roller in a paper making machine that impresses a watermark into the paper

densitometer an electronic instrument used to measure the quantitative colours or density of colour in printing

design the process of solving problems by creating ideas to come up with a solution

design ideas a wide range of different ideas to solve a problem. Often communicated in sketch form

designers people who are employed to create design ideas for a client

developing ideas the process of taking an idea and improving / modifying it so that it is the best solution to a problem

digital camera a device that takes images and records them digitally on computer disc or other device

division of labour when a crafts person is employed to make only part of the overall product

drafting aid any tool which helps to reduce the difficulty in reproducing lines and curves on paper

die cutting a tool similar to a pastry cutter using thin blades pressing down to cut and / or crease irregular shapes in card

electrolytic process when an electric current is passed through a material (electrolyte) via a cathode and anode releasing an element held within the electrolyte

embossed a process which raises the surface of a material using a press or stamp. Normally used to give an interesting visual affect

EPOS Electronic Point of Sale

ergonomics the study of how products and environments are designed to be efficient for human users

flexography a printing process which uses flexible printing plates, usually rubber or plastic. Used predominantly for packaging

flow chart a chart using symbols to show the sequence of a process

FMS Flexible Manufacturing System

formative review evaluation which is on going during the design and make processes – often in the form of annotation

Fourdrinier a machine used to convert wood pulp into paper using a series of rollers, pressers and dryers

function the purpose of a product – what it does

Gantt charts a chart to show how a number of tasks or processes are planned to be completed in a given time, often concurrently

graphic identity when an organization uses visual symbols or logos to separate them from other companies

gravure a method of four-colour printing using the intaglio process. Normally for high volume runs

hot foil blocking a method of transferring a thin foil coating to paper or card for visual affect

ICT Information Communication Technology

injection moulding a process where molten thermoplastic is injected under high pressure into a die cavity

innovative to introduce new methods, ideas that are different from what has already been attempted

input information or material at the start of a system i.e., paper and ink at the start of print production

Internet a global network of computers allowing access to a huge amount of information

ISDN Integrated Services Digital Network

isometric a method of drawing in 3D using 30° axes

ISP Internet Service Provider

JIT 'just-in-time' – meaning the delivery of materials/components to the manufacturing area only as they are needed

JPEG Joint Photographic Experts Group. A form a compressed computer file used for images published on the internet

Kraft papers strong brown paper made from sulphate pulp. Used for backing books

laid paper paper showing the wire marks of the mould or dandy roll used in manufacture

lay planning the method of placing a pattern or net onto sheet material to minimize waste

letterpress the original 'relief' method of printing whereby a raised image or letter is primed with ink before paper is pressed onto it

lithography a printing process that produces an image from a flat dampened plate using greasy ink, based on the principle that water and oil do not mix

manufacturing cells a production system that incorporates a number of people and machines working together, being responsible for what is produced

marketing the selling of a service or product to a customer

mass production the production of a component or product in large numbers

mechanical a term describing a process using machines or a system using mechanisms, or a person skilled in using machinery

mechanization when a system or process is converted from being operated by hand to machine

miniaturization to make small. A result of the desire for 'small is beautiful' philosophy

mock-up a model/replica of a design in 3D used for evaluation and testing

mood boards a collage of images related to a topic or theme. Used as a springboard for designing

mould a hollow or convex shape used for shaping a material, like in vacuum forming

net the 2D shape or development of a 3D product in sheet form

output the product of a system from input and process

oxygen furnace used in steel manufacture to create high temperatures when smelting

parison see 'blow moulding'

perspective a form a drawing in 3D using vanishing points along a horizontal axis or eye line

PICT file abbreviation for picture format, used in computers for saving pictures

plotter/plotter cutter an computer controlled output device for producing accurate lines or cuts on card or paper

point-of-sale graphic products which are used to advertise a product in a shop, often mounted on a counter or on a shelf

portfolio the presentation of design work

presentation drawings drawings used to communicate a design in a suitable form for the client

product disassembly critical analysis of a product, breaking it down into it's individual elements

product models accurate replicas of a design in 3D for evaluation in the marketplace

production cells similar to manufacturing cells

proof copy a representation on paper of the printed product to check the progress and accuracy of the work

prototype a pre-production version of a product, used to check whether it meets requirements

pulp mill where wood is broken down into fibres and lignin for paper making

pure metal a metal without impurities – a single element

quality assurance a policy or procedure written to ensure that a product reaches the customer to the correct specification

quality control systems put in place to check quality during manufacture – i.e. gauges, visual checks, etc.

questionnaire a series of open or closed questions used in a survey to find out peoples views on an issue

remote manufacturing where a design is sent to another place to be manufactured

scale of production the type of production – batch, mass, etc.

scanner an input device used to import images into a computer for DTP

scoring the process of putting creases onto card for folding

screen printing a low volume method of printing onto different surfaces using templates, a wire mesh and ink

seasoning the process of converting timber into a usable form by drying in air or kiln

shrink wrapping where a product is wrapped in plastic sheet which is then heated to shrink tightly around it. Use as protective packaging

specification a set of clearly indicated criteria that the final solution must meet

sprue either a channel through which metal or plastic is poured into a mould or the metal or plastic which has solidified in a sprue; often holds pieces in model kits

stereolithography 3D modelling using lasers to solidify liquid plastics. Complex shapes can be produced – often called rapid prototyping

storyboard where a series of stages in design or making are shown using pictures, like in a cartoon strip

summative review the critical evaluation of a task or product after it is completed to see if it has met the design specification

thermoplastic a type of plastic which softens under heat and can be re softened many times

three dimensional (3D) modelling a method of communicating a design in three dimensions, either by using suitable resistant materials – card, wood, MDF, clay etc., or by using computer software to show virtual 3D images

thermosetting a type of plastic which once set cannot be re-softened or melted

TIFF Tagged Image File. A type of file saving format for images in DTP

tolerance the upper and lower limits of a dimension, i.e. 25mm +/-0.5mm = 24.5/25.5mm

TQM Total Quality Management – where the workforce are given responsibility for ensuring quality by using their skills and ideas to continually improve quality

trade mark a company's recognized trade symbol, logo or typeface which no other company can use

typeface a type design (text style) – including variations like italic and bold

uncoated refers to paper used in books, catalogues, etc. in a range of finishes, i.e. rough, calendered (smooth), etc.

vacuum forming a process using thin plastic sheet which is formed around a mould using atmospheric pressure – used in blister packaging

veneers thin strips of hardwood (0.5mm thick) used for laminating onto cheap material to improve visual appearance

virtual design designs created totally on a computer in 3D

watermark a distinctive design incorporated into a paper during the manufacturing process

weeding a process where a chemical is applied to a vinyl sticker to allow the background to be removed leaving only the desired graphic

wood pulp the material created when raw wood is processed to separate the fibres from the lignin. Used in paper making

wt method by which a component is controlled through the production cycle in relation to other components

wove paper paper with a smooth, even and fine surface

Index